New Directions in Philosophy and Cognitive Science

Series Editor
Michelle Maiese
Emmanuel College
Boston, MA, USA

This series brings together work that takes cognitive science in new directions. For many years, philosophical contributions to the field of cognitive science came primarily from theorists with commitments to physical reductionism, neurocentrism, and a representationalist model of the mind. However, over the last two decades, a rich literature that challenges these traditional views has emerged. According to so-called '4E' approaches, the mind is embodied, embedded, enactive, and extended. Cognition, emotion, and consciousness are not best understood as comprised of brain-bound representational mechanisms, but rather as dynamic, embodied, action-oriented processes that sometimes extend beyond the human body. Such work often draws from phenomenology and dynamic systems theory to rethink the nature of cognition, characterizing it in terms of the embodied activity of an affectively attuned organism embedded in its social world. In recent years, theorists have begun to utilize 4E approaches to investigate questions in philosophy of psychiatry, moral psychology, ethics, and political philosophy. To foster this growing interest in rethinking traditional philosophical notions of cognition using phenomenology, dynamic systems theory, and 4E approaches, we dedicate this series to "New Directions in Philosophy and Cognitive Science."

If you are interested in the series or wish to submit a proposal, please contact Amy Invernizzi, amy.invernizzi@palgrave-usa.com.

Miguel A. Sepúlveda-Pedro

Enactive Cognition in Place

Sense-Making as the Development of Ecological Norms

palgrave
macmillan

Miguel A. Sepúlveda-Pedro
Institute for Philosophical Research, National Autonomous University of Mexico
Mexico City, Mexico

New Directions in Philosophy and Cognitive Science
ISBN 978-3-031-20281-0 ISBN 978-3-031-20282-7 (eBook)
https://doi.org/10.1007/978-3-031-20282-7

This Palgrave Macmillan imprint is published by the registered company Springer Nature Switzerland AG.
The registered company address is: Gewerbestrasse 11, 6330 Cham, Switzerland

To Brenda

Acknowledgments

Any human achievement is the result of collective action. For this reason, I will be eternally thankful to many people that supported and helped me to materialize this work.

This work was supported by UNAM Posdoctoral Program (POSDOC). I must also recognize the effort of the working-class people of Mexico. Their labor paid for the scholarship I obtained from the National Council for Science and Technology of Mexico (CONACYT) to accomplish my doctoral research at the University of Montreal.

I am also profoundly grateful to Maxime Doyon and David Morris, my PhD advisors, for all their teachings. They did not only assist me in finding the path of a professional researcher in philosophy, but their reviews and comments on earlier versions of the book manuscript were also crucial to building more rigorously the philosophical augments I expose here.

Ezequiel Di Paolo's review and comments were also highly constructive in refining and clarifying my arguments about the complex relationship between the enactive approach and phenomenology.

I thank Giovanni Rolla whose comments were precious to putting the final version of the book manuscript in shape.

Over the past ten years, I also had many fruitful conversations that improved my understanding of the proposals of enactivism, phenomenology, and ecological psychology. I can only mention some of the people with whom I had the most significant ones: Manuel Heras- Escribano, Tom Froese, Shaun Gallagher, Tony Chemero, Eric Rietveld, Scott Marratto, Xavier Barandiaran, Dan Hutto, Sergio Martínez, Ricardo

Vázquez, Juan González, Maxwell Ramstead, Laura Mojica, Susana Ramírez-Vizcaya, Jelle Bruineberg, and Alejandro Macías.

I also appreciate the effort of my writing editor, Rohan Quinby, who was always worried about my requirements and helped me become a more solid academic writer.

My dear friend, Ivonne E. Kuri, sadly died three years ago. She significantly influenced my philosophical career. I will be forever thankful to her.

I'll be always in debt to my parents, Judith Pedro and Miguel Sepúlveda-Torres, to my sister, Karla Sepúlveda, and more especially to my wife, Gina López. They have always been there, by my side, supporting and helping me to overcome all the challenges I have met in my professional career.

Contents

List of Figures

List of Tables

CHAPTER 1

Introduction: From the Embodied Mind to the Emplacement of the Living Body

What is the role of the environment when we perceive, act, imagine, remember, think, and interact with other living beings? Is the environment a bare material landscape that provides information to our senses? Is it nothing more than an arena, or a background upon which our everyday lives occur? Is the environment nothing more than a source of resources affording living organisms opportunities for survival, while other times threatening them with dangers? Or is the environment also a web, one in which all living beings are presently and historically interconnected? Is the environment also something meaningful, a realm where the values of human and animal societies are sedimented? Does the activity of all living organisms modify the environment? Do we share and transform the history of the world around us by the mere fact of being alive? The answers to all these questions depend on our preconceptions about what we are as living, and as minded beings. Contemporary cognitive science offers at least two different responses to these questions, stemming from two divergent visions of mind and life within the field.

The traditional view, which I will call *brain-centered cognitive science*, conceives the mind as an abstract computational system that is physically implemented in the human brain, and sees cognition as a function of information processing. The body is, from this perspective, a bundle of sensory inputs that provide us with information about our surroundings, and a set of motor outputs that execute the actions our brain commands.

M. A. Sepúlveda-Pedro, *Enactive Cognition in Place*, New Directions in Philosophy and Cognitive Science,
https://doi.org/10.1007/978-3-031-20282-7_1

The energetic and material properties of the things we observe entirely determine the nature of our surroundings. The material environment provides information directly to our sensory organs, or in our day, through our instruments of scientific observation. We can and do modify the environment, thanks to our actions and the ways that we manipulate things around us, but this is only insofar as *we* cause material changes. As we should know by now, altering the environment can cause unpredictable consequences for the survival of both our own and other species. In the present, like in the past, the material conditions of the environment put pressure on the lineages of organisms, and shape the phylogenesis of all species. From this brain-centered standpoint of cognitive science, therefore, the environment is simply the material basis that predetermines the contents of our cognitive activities, and causally shapes the evolution of our cognitive apparatus.

The opposite view within cognitive science tells us something different. For those who theorize what I will call a *world-involving cognition*, our minds are embodied in dynamic self-organizing systems that reach across the brain, the body, and the environment. Cognition, in this view, involves processes that evolve in time, and in accordance with the ever-changing conditions of these extended systems. From this standpoint, our surroundings are not entirely determined by their simple materiality, because the bodily activity of an embodied mind continuously alters and shapes—both materially and meaningfully—the configuration of the environment. Our minds are thus embodied and deeply entangled with the world around us. Does that mean the environment is co-determined with our cognitive lives? More broadly, does the environment coevolve with the living organisms that inhabit it?

This book aims to enrich our understanding of the role the environment plays from the perspective of world-involving cognition and attempts to offer some clues in response to some of the questions I have asked above. More specifically, my purpose in *Enactive Cognition in Place* is to uphold the claims of what has become known in cognitive science as the *enactive approach* (Di Paolo et al., 2017, 2018; Thompson, 2007; Varela et al., 2016). More commonly known as autopoietic enactivism, it is a view that sees cognition as emerging from the organizational nature of the living body, and evolving through sensorimotor, social, and linguistic interactions with both our natural and sociocultural surroundings. I want this book to help enlarge our understanding of how living bodies are ecologically situated in places, and further, to see that these places are themselves developmentally and historically constituted by the active participation of the living populations they host.

By way of an introduction, this chapter will briefly survey the central claims of the enactive approach. But it will also argue the *necessity* of uncovering the roots of cognition, not only in the nature of our living bodies, but also in the ecological places we live in, in accordance with this perspective.

1.1 The Enactive Approach: Biological Autonomy and Sense-making

From the last decade of the twentieth century to now, the enactive approach has sought to restructure our scientific understanding of cognition by rejecting the main assumptions of the most traditional—and still predominant—view in cognitive science today (Varela et al., 2016). I call this prevailing view brain-centered cognitive science because it locates all the relevant causal processes underlying human cognition almost exclusively in the brain. The localization of cognition in the confines of the head is tied to the claim that cognition is a computational function of information processing, one that serves to solve the problems we face in our natural and cultural environments. These computational models of the mind have helped explain many cognitive and brain processes, particularly the most abstract forms of cognition, such as reasoning and mathematical thinking (Thagard, 2005). At the same time, they have stumbled when forced to describe our more basic cognitive functions, such as perception and intelligent behavior.

When researchers within brain-centered cognitive science encountered difficulties explaining these basic cognitive functions, a new generation of cognitive scientists and philosophers suggested that our explanations of cognition needed to incorporate relevant bodily and environmental aspects that help shape and constitute minds. Historically, the use of models of dynamical systems theory was a very effective mathematical tool for describing cognition as an interactive process involving not only the brain, but also the body and the environment (van Gelder & Port, 1995). The empirical success of these models, especially in the field of artificial intelligence (Brooks, 1999), opened the door for the emergence of a new kind of cognitive science, which was initially labeled *embodied cognition* (Shapiro, 2010).

Embodied cognition has never been a unitary research program. Instead, its various approaches tend to focus on different bodily and environmental aspects that can influence and even constitute cognitive processes. The philosophical and scientific backgrounds of each type of embodied cognition are often divergent and can even be contradictory. Indeed, the opposition of embodied cognition to the dogmas of brain-centered cognitive science significantly differs in its degree of radicality. For this reason, I prefer to use the label world-involving cognition (Di Paolo et al., 2017) to name those views standing in radical opposition to brain-centered cognitive science within embodied cognition. Such a label implies the rejection of a fundamental assumption held by the neurocentric stance: the problematic gap between the mind and the world.

Because of its commitment to a world-involving cognition, the enactive approach has been one of the most radical antagonists of traditional brain-centered cognitive science. In particular, this enactivism not only denies that cognition is enclosed in the boundaries of the head, but also rejects the claim that cognition consists in information processing. For the enactive approach, cognition is best understood as a form of what its supporters name *sense-making* (Di Paolo & Thompson, 2014).

The concept of sense-making is critical to overcoming the traditional gap between mind and world. Sense-making describes the co-emergence and co-determination of the self-identity of a living and cognitive agent, and the correlated meaningful world that is relevant to its activities (Thompson, 2007). In this sense, the environment we interact with is not entirely determined by its material constitution. It is also constituted by the meanings that living organisms give to their actions.

The notion of sense-making needs to be understood in the framework of the enactive *theory of biological autonomy* (Weber & Varela, 2002). This theory argues that the physical constitution and behavior of living systems cannot be entirely determined by causes external to their dynamic constitution, as is the case with nonliving systems. All physical systems, from galaxies to stones, tend to disintegrate their systemic organization and cease to exist as the organized systems we may observe (i.e., as galaxies and stones). This tendency is called *entropy*. Living systems, on the other hand, can temporarily avoid entropy by interacting adaptively with the environment (Schrodinger, 2012). For the enactive approach, living systems resist entropy thanks to their *autonomous organization* (Barandiaran & Moreno, 2006). Organisms are constituted by a network of interdependent processes that make these physical systems partially independent from the causal processes outside the network (Di Paolo & Thompson, 2014). Nonetheless, the autonomy of living systems depends on the capacity of these systems to interact adaptively with their surroundings to maintain their organizational viability (Di Paolo, 2005). Living systems are essentially *self-maintaining systems*.

The paradigmatic example of biological autonomy is the cell (Thompson, 2007). The interdependent processes that take place in a cell's interior help it constantly regenerate its chemical composition and sustain the membrane that separates its internal organs from the rest of the environment. To maintain this network, however, the cell must also interact with the environment, exchanging the matter and energy necessary to renovate its chemical components. The cell can partially adapt its internal composition, as well as its interactions with its surroundings in order to maintain its metabolism.

The autonomy of living systems entails that organisms behave according to norms enacted by their own systemic organization (Di Paolo & Thompson, 2014). Organisms possess what enactivists call intrinsic normativity, since the purpose and guidance of their individual development and behavior are determined by their self-maintenance activity (Mojica, 2021). As a result of their biological autonomy, organisms are sensitive to specific aspects of their surroundings. The environment is disclosed as a realm of possibilities for action that can help or threaten an organism's self-maintenance. The disclosure of the environment according to norms that are enacted by autonomous systems is what constitutes sense-making (Thompson, 2007).

At the biological level, metabolic needs form the basis of sense-making norms. All organisms normatively select and interact with features of the environment that are good or bad for the metabolic activity of their bodies. Nonetheless, cognition as such begins at the sensorimotor level of autonomy and sense-making (Barandiaran, 2017).

In sensorimotor interactions like walking and swimming, the body self-organizes to sustain a network of interdependent processes, such as in the coordination of legs and arms, and these processes can acquire an autonomy of their own, thereby becoming a sensorimotor habit (Di Paolo et al., 2017). When such a thing happens, the living body follows a new set of norms that function to preserve the recurrent organization of the extended sensorimotor agent-environment system. That is, the purpose of sustaining habitual interactions with the environment guides the activity of the living body, and the environment is perceived as an opportunity to perform these actions.

Sensorimotor interactions with the environment often involve interactions with other living agents. These interactions can produce new and emergent self-organizational processes between two or more organisms, thereby establishing new levels of autonomy and normativity for the participants of such an intercorporeal coordination. The enactive approach calls this emergent level of normativity *participatory sense-making* (De Jaegher & Di Paolo, 2007). Practices like fighting, dancing, and playing sports are just some examples of bodily social interactions that can lead to this intercorporeal form of sense-making.

For the enactive approach, interactions between living bodies form the basis of human sociocultural and linguistic practices (Di Paolo et al., 2018). The ability of human social groups to use recurrent and successful utterances (either as gestures or as verbal expressions) to co-regulate their bodily interactions, the establishment of hierarchies in which individual members co-regulate these interactions, the creation of dialogic structures to explicitly modify and manipulate social practices, and the possibility of internalizing dialogic structures to self-modulate individual bodily actions are some of the linguistic resources enabling the emergence of abstract forms of human cognition (Di Paolo, 2016a). The complete picture of how concrete sensorimotor interactions become abstract forms of cognition is an ongoing research project for the enactive approach, although the basic understanding of the path from life to language is already sketched out.

In sum, for the enactive approach, cognition is seen as grounded in sensorimotor activity, but only insofar as the constitution of all sensorimotor activity has its origins in the autonomy of life. There is thus a deep continuity between the processes of mind and life (Thompson, 2007), and since life and cognition depend on the irreducible systemic unity of the agent–environment coupling, we can only understand processes of sense-making and cognition through this unity.

1.2 The Missing Ecological Dimension of Sense-making

It is safe to say that the enactive approach is now one of the most promising branches of the study of world-involving cognition. It has extended the study of cognition to all of its bodily, affective, phenomenal, interactive, and linguistics dimensions. Its methodologies include first-, second-, and third-person perspectives (Di Paolo et al., 2018). The application of its central ideas can be seen in the development of artificial life, embodied robotics, psychopathology, disability and rehabilitation, education and pedagogy, musical cognition, dance performance, architecture, archaeology, anthropology, sport sciences, ethics, and more (see Di Paolo, 2016b). Despite this success, it remains true that the study of such a complex phenomenon as cognition will always require improvements and rectifications in any research program's theoretical and empirical tools.

One of the main challenges for the enactive approach is the proper recognition and description of all the constitutive aspects of cognition arising from the agent–environment coupling. Indeed, one of the critiques leveled against the enactive approach is that there is a tendency among its supporters to neglect the role the environment plays in the constitution of cognition and sense-making, particularly when they miss its function as a normative ecological field (see, e.g., McGann, 2014).

In the main works of the enactive approach, we find two sorts of references to the environment that surrounds a living body. The first type of reference understands the environment as the raw, material surroundings of an agent. This material environment is eventually transformed into a meaningful world when an autonomous system emerges and interacts normatively with its material environment. The act of constitution of the autonomous system and the environment as a meaningful world occurs precisely in sense-making. A paradigmatic example of sense-making and the dual character of the environment in enactive approach literature is that of *Escherichia coli* bacteria, and their tendency to swim towards glucose. These bacteria are said to find a (neutral value) chemical compound that *becomes* a nutrient (significant or value-laden) thanks to their metabolic needs (Thompson, 2007).

The understanding of the environment in its duality points to the innovative idea that a meaningful world emerges for living agents as a consequence of the autonomous organization of the living body. However, this description also neglects the active role the environment plays in acts of sense-making and cognition. The environment a living agent encounters at any point in its existence is not a bare material landscape, but a normative and active field that places constraints and pressures on the activity of that living agent. That is to say that normative—and not simply bare material—constraints shape an agent's development and enable their norms of interaction with the environment.

Good examples of normative environmental constraints happen in human sociocultural environments. There, social norms shape our behavior. Doing sports or playing music are clear cases where we need to incorporate a normativity that is not established by our agency, but by the implicit or explicit agreements of a community around how to accomplish these skillful practices successfully (Rietveld, 2008). But, normative constraints do not only originate synchronically with others in sociocultural practices. There is also a fundamental diachronic, or historical dimension of normative constraints that are sedimented in the proximate environments we inhabit.

We cannot enact new norms of sensorimotor habits unless we have previously embodied other sensorimotor norms. A child cannot run if she cannot first learn to stand up. I cannot play soccer if I have not first learned to walk and run. The enactive approach account of sensorimotor habits implicitly acknowledged this phenomenon (Di Paolo et al., 2014). The description of sensorimotor habits focuses on the constitution of the autonomous organizations that constitute these new networks of bodily organization. Nevertheless, the sensorimotor body also depends on the materiality of the environment as much as the living body does for its vital needs. Moreover, the environment embodies the acquisition of sensorimotor norms.

Learning a new bodily skill like swimming changes not only the organization of my body, but also the way the water appears to me as an opportunity to move in without drowning. Indeed, enacting sense-making norms for swimming not only reveals a meaningful world—a water world—to the agent, but also affords the possibility of enacting a whole new developmental pathway. Once we start swimming, we can potentially begin diving, dancing, relaxing, or performing an endurance sport in the water. The evolution of these practices will depend on many aspects of our lives. But if we focus on the sensorimotor component of the practice, we can see that if we do not learn to swim "properly," we cannot unfold a specific developmental pathway. Each swimmer can become a good performer in the diverse practices of water sports, among other things, according to an individual's sensorimotor development. The water world does not appear to all individuals in the same way. The way the water world will enable some pathways instead of others for a single individual will significantly depend on her sensorimotor development. At the same time, normative constraints of the practice co-enacted with others also co-determine how an agent discloses the world. Different ways of teaching and learning to swim around the world deliver different results in the performance of swimming. Therefore, the constitution of a meaningful environment implies not only the emergence of a world of opportunities for action, but also the emergence of different developmental pathways that shape the behavior, the body, and the environment of agents at many temporal scales (cf. Chemero, 2009).

This logic of development does not only occur among different levels of sensorimotor habits. Previously established biological norms also constrain the emergence of sensorimotor norms. When jogging, for instance, we need to adapt our practice to the limits of our cardiovascular and respiratory performance. We can improve our endurance through training, and the acquired sensorimotor norms can reshape our biological norms, but this process also involves development. A well-trained jogger also discloses the environment differently, and the possibilities for her development are also different. Material and normative constraints, sedimented in training places, can undoubtedly lead to remarkable differences.

At the biological level itself, ontogenetic and phylogenetic normative constraints are also central to shaping sense-making pathways. Depending on the kind of environment, an organism living in that environment might express one type of gene instead of another (Barandiaran & Egbert, 2014). While the environmental conditions that alter gene expression are often material, they can also be normative (Walsh, 2018). For instance, interactions with parental organisms, predator–prey interactions, as well as individual development of survival skills all sediment in the environment a set of constraints. These constraints alter the biological development of an organism, and eventually the evolution of whole species, increasing or decreasing their chances of survival and reproduction.

That both synchronic and diachronic dimensions of normative constraints are sedimented in the environment is relevant to the study of cognition within the enactive approach. It helps us recognize, for example, the emergence of concrete differences in the historical development of individuals, sociocultural communities, and species. It also allows us to understand sense-making in its historical dimension as a process of transformation, where living bodies actively participate in shaping the pathways of their development, and in the historical transformations of their communities, and in the places they live. Acknowledging the ecological and normative dimensions of the environment helps us to rethink what sense-making consists of. For if sense-making does not occur in an abstract, purely material environment—one where a self-identity and a meaningful world simply emerge—but in a normative field where the ecological bond of the body and the environment is already present, then we can see that sense-making occurs as an act of transformation and development.

To help us rethink the role the environment plays as a normative ecological field in processes of life and cognition, and to try to redefine the concept of sense-making as a process of development, I propose to incorporate three main theses into the theoretical body of the enactive approach: (1) sense-making consists in processes of norm development; (2) the environment agents encounter is an active ecological field; and (3) the actual enactment of sense-making norms occurs in a domain consisting of multiple normative dimensions that I will call enactive place. I have derived these ideas from conceptual resources provided by Merleau-Ponty's phenomenology (2012), ecological approaches to cognition based on Gibson's (2015) ecological approach to perception, and from Edward Casey's (1993) philosophy of place. In what remains of this chapter, I will outline the plan for the rest of the book and show how I will incorporate these three ideas into the enactive approach's theoretical corpus.

1.3 Book Outline

Chapter 2 will introduce us to the divergent paths of cognitive science as we currently find them. First, I will mark out the division between brain-centered cognitive science and embodied cognition, and briefly survey the general assumptions of the three primary forms of brain-centered cognitive science: *cognitivism*, *connectionism*, and *predictive coding*. We will have cause to find shortcomings in each of these forms, and I will examine some scientific and philosophical problems that result from enclosing the mind inside the boundary of the head. The chapter will also sketch out differences in the field of embodied cognition, particularly in relation to their opposition to the central dogmas of brain-centered cognitive science. Finally, I will locate the enactive approach among the most radical forms of opposition to brain-centered cognitive science.

Chapter 3 analyzes the main claims of the enactive approach, illuminating the radical opposition of this approach, not only to the explicit dogmas of brain-centered cognitive science, but also to the epistemological and ontological bases of these dogmas. Hand in hand with phenomenology, the enactive approach distinguishes itself from any other form of enactive cognition. Although the enactive approach inaugurated this branch of cognitive science, other forms of enactive cognition such as sensorimotor enactivism and radical enactivism have emerged and count some successes to their credit. Nonetheless, I will argue these other enactivist proposals are simply weak forms of enactivism because they fail to break with the problematic philosophical and scientific presuppositions of brain-centered cognitive science. An enactive approach that postulates concepts of biological autonomy and sense-making is strong because it radically refuses the scientific and philosophical assumptions of brain-centered cognitive science. Ultimately, I argue that the enactive approach leads world-involving cognition to a more radical transformation of our view of mind and life than any other form of cognitive science.

In Chap. 4, I will examine the problems caused by the traditional descriptions of sense-making, and of the dual nature of the environment as both a material landscape and a meaningful world. I argue against critics of the enactive approach that the problem of the duality of the environment is not an ontological or epistemological dualism as plagues brain-centered cognitive science. Rather, the environmental dualism favored by the enactive approach is phenomenological and does not contradict a view of world-involving cognition. However, I also show why the current description of sense-making found in the enactive approach is limited, because it neglects the body-world entanglement. This entanglement names the ecological bond that anticipates any act of sense-making. I will also argue that such entanglement leads us to think of sense-making as a process of norm development and points to how Merleau-Ponty's phenomenology sheds light on this issue. At the end of this chapter, I will support the claim that the body–world entanglement is a precondition for sense-making in the same way that perception, from a phenomenological analysis, cannot occur unless there is a previous layer of connection between the body and the world.

Once we have considered the body–world entanglement, as well as the claim that sense-making consists in norm development, in Chap. 5 I propose to show that the environment has a more active role than the enactive approach has recognized thus far. Agents establish an asymmetric relationship with the environment because the agent modulates and transforms this relationship, but that does not mean the environment is a bare set of material constraints. The environment is rather an ecological field of normative forces that shape the developmental pathways of the whole agent-environment system. It is more helpful to describe the environment as a field of affordances, as some scholars of Gibson's ecological approach to perception have already done.

However, we need to be clear that in contrast to this ecological view, the constitution of a field of affordances implies a process of norm enactment that individuals and other organisms realize during their lives. Consequently, we cannot assume that a field of affordances is pre-given due to causal processes external to the agent–environment system development. Instead, following Merleau-Ponty's phenomenological account of spatial levels and shifts of spatial levels, I propose thinking about the environment's constitution as an enactive-ecological field. Accounts of how new perceptual meanings emerge from bases provided by already established spatial levels, discloses the already existent bond between the body and the world. This view is compatible with the ecological claims that the environment we deal with is a normative field of affordances but assumes that such a field is enacted in the developmental pathways of the agent–environment system.

If we accept that sense-making consists in norm development, and that the environment is an enactive-ecological field shaping the developmental pathways of the whole agent–environment system, we can understand the real situation of agents, and exactly where sense-making occurs. This is to say, as I will argue in Chap. 6, sense-making occurs in places. I recover Casey's descriptions of place to sketch an enactive account of place. In such a definition of place, multiple ecological normativities converge and constrain the developmental pathways of agents. Simple materiality does not entirely determine places; they are also constituted by the interaction of multiple agents in a material medium that incarnates the social lives of communities. Although places are more easily discernible in the sociocultural practices of human groups, I utilize aspects of Merleau-Ponty's philosophy of nature to theorize that communities of organisms of different species also constitute communities, ones that coevolve historically with the material landscapes these organisms inhabit.

In the final chapter, I will examine some of the consequences of studying cognition in place from an enactivist perspective. The notion of enactive place challenges the study of cognition because it shows us the overwhelming complexity of the ecological and normative web of life and cognition. But all is not lost, for such a challenge can help us reflect on the methodologies and aims of enactive cognition, as a scientific and philosophical research program, and strengthen calls to study cognition in a more situated way.

References

Barandiaran, X. E. (2017). Autonomy and enactivism: Towards a theory of sensorimotor autonomous agency. *Topoi, 36*(3), 409–430.

Barandiaran, X. E., & Egbert, M. D. (2014). Norm-establishing and norm-following in autonomous agency. *Artificial Life, 20*(1), 5–28.

Barandiaran, X. E., & Moreno, A. (2006). On what makes certain dynamical systems cognitive: A minimally cognitive organization program. *Adaptive Behavior, 14*(2), 171–185.

Brooks, R. A. (1999). *Cambrian intelligence: The early history of the new AI.* MIT Press.

Casey, E. S. (1993). *Getting back into place: Toward a renewed understanding of the place-world.* Indiana University Press.

Chemero, A. (2009). *Radical embodied cognitive science.* MIT Press.

De Jaegher, H., & Di Paolo, E. A. (2007). Participatory sense-making. *Phenomenology and the Cognitive Sciences, 6*(4), 485–507. https://doi.org/10.1007/s11097-007-9076-9

Di Paolo, E. A. (2005). Autopoiesis, adaptivity, teleology, agency. *Phenomenology and the Cognitive Sciences, 4*(4), 429–452. https://doi.org/10.1007/s11097-005-9002-y

Di Paolo, E. A. (2016a). Participatory object perception. *Journal of Consciousness Studies, 23*(5–6), 228–258.

Di Paolo, E. A. (2016b). Enactivismo. In C. E. Vanney, I. Silva, & J. F. Franck (Eds.), *Diccionario Interdisciplinar Austral.*

Di Paolo, E. A., Barandiaran, X. E., Beaton, M., & Buhrmann, T. (2014). Learning to perceive in the sensorimotor approach: Piaget's theory of equilibration interpreted dynamically. *Frontiers in Human Neuroscience, 8*, 551.

Di Paolo, E. A., Buhrmann, T., & Barandiaran, X. E. (2017). *Sensorimotor life: An enactive proposal.* Oxford University Press.

Di Paolo, E. A., Cuffari, E., & De Jaegher, H. (2018). *Linguistic bodies: The continuity between life and language.* MIT Press.

Di Paolo, E. A., & Thompson, E. (2014). The enactive approach. In L. Shapiro (Ed.), *The Routledge handbook of embodied cognition* (pp. 86–96). Routledge.

Gibson, J. J. (2015). *The ecological approach to visual perception.* Psychology Press.

McGann, M. (2014). Situated agency: The normative medium of human action. *Synthesis Philosophica, 58*(2), 217–233.

Merleau-Ponty, M. (2012). *Phenomenology of perception.* (D. Landes, Trans.). New York, NY: Routledge.

Mojica, L. (2021). The enactive naturalization of normativity: From self-maintenance to situated interactions. *History and Philosophy of the Life Sciences, 43*(4), 1–27.

Rietveld, E. (2008). Situated normativity: The normative aspect of embodied cognition in unreflective action. *Mind, 117*(468), 973–1001.

Schrodinger, E. (2012). *What is life?: With mind and matter and autobiographical sketches.* Cambridge University Press.

Shapiro, L. A. (2010). *Embodied cognition.* Taylor & Francis.

Thagard, P. (2005). *Mind: Introduction to cognitive science* (2nd ed.). MIT Press.

Thompson, E. (2007). *Mind in life: Biology, phenomenology, and the sciences of mind.* Harvard University Press.

van Gelder, T., & Port, R. F. (1995). It's about time: An overview of the dynamical approach to cognition. In R. F. Port & T. van Gelder (Eds.), *Mind as motion: Explorations in the dynamics of cognition* (pp. 1–43). MIT Press.

Varela, F. J., Thompson, E., & Rosch, E. (2016). *The embodied mind: Cognitive science and human experience* (2nd ed.). MIT Press.

Walsh, D. M. (2018). Objectcy and agency: Towards a methodological vitalism. In J. Dupré & D. J. Nicholson (Eds.), *Everything flows: Towards a processual philosophy of biology* (pp. 167–185). Oxford University Press.

Weber, A., & Varela, F. J. (2002). Life after Kant: Natural purposes and the autopoietic foundations of biological individuality. *Phenomenology and the Cognitive Sciences, 1*(2), 97–125. https://doi.org/10.1023/a:1020368120174

CHAPTER 2

Worlds Apart: Are We Enclosed Inside Our Heads?

Cognitive science is a multidisciplinary field that looks for natural explanations of human activities such as perception, motor action, memory, imagination, planning, and reasoning. We call these activities cognition because they procure knowledge about the conditions of the environment and make our intelligent actions possible. Currently, there are two main perspectives found within cognitive science. The first we might call *brain-centered* cognitive science because it locates all the relevant causes of cognition in brain processes. The second one we can call world-involving cognition because it maintains that bodily and environmental processes are fundamental to the constitution and transformation of our cognitive functions.

This chapter begins by demonstrating that all forms of brain-centered cognitive science share two central ideas: *neurocentrism* and *computationalism* (Sect. 2.1). The synthesis of these two ideas has tended to predominate in many scientific explanations of cognition. However, despite the success of these explanations, they share a key and problematic assumption: the mind–world dichotomy (Sect. 2.2). The idea of a mind–world dichotomy is an inheritance from the Empiricist and Cartesian philosophical traditions and is at the root of multiple philosophical and methodological issues when it comes to the study of the mind (Sect. 2.3). In light of these difficulties, we can turn to alternative views of embodied cognition that offer proposals to solve some problems of brain-centered

M. A. Sepúlveda-Pedro, *Enactive Cognition in Place*, New Directions in Philosophy and Cognitive Science,
https://doi.org/10.1007/978-3-031-20282-7_2

cognitive science, often by including certain bodily and environmental processes as fundamental for the causal constitution of cognition. However, I will argue that only the radical forms of embodied cognition that I name world-involving cognition can genuinely overcome the problematic mind–world dichotomy and found cognitive science on a whole new basis (Sect. 2.4).

2.1 Brain-Centered Cognitive Science

A widely accepted view among the contemporary community of cognitive scientists is that cognition is like a function of information processing accomplished by a computational system. The mind, they hold, amounts to the characteristics of the software we run on our personal computers (Thagard, 2005). From this standpoint, the mental software of human beings is physically implemented in the sophisticated machinery of the brain, and the brain is the hardware that runs the mental software. I call the claim that cognition consists of information processing *computationalism* and the belief that cognition is physically implemented in the brain *neurocentrism*.

In this section, I briefly review the origin of the synthesis of computationalism and neurocentrism in cognitivism (Sect. 2.1.1) and the subsequent adoption of the same synthesis in the explanatory frameworks of connectionism (Sect. 2.1.2) and predictive processing (Sect. 2.1.3).

2.1.1 Cognitivism

Cognitivism is the most classical and paradigmatic form of brain-centered cognitive science. This scientific account of mental phenomena developed the metaphor of the mind functioning as a computational system and the brain working as a computer. The rise of computational science in the middle of the twentieth century prompted scientists and philosophers to use computational models to produce well-founded hypotheses about the formal processes at play in cognition, notably in human reasoning (e.g., Newell & Simon, 1961). Since these first computational models were also valuable in explaining brain mechanisms at the macroscale, the isomorphism between mental and brain functions suggested that the brain was the physical implementation of the mind (McCulloch, 1949).

Computational systems accomplish information-processing tasks that transform discrete values or inputs into other values or *outputs* thanks to applying specific rules or *algorithms*. The hypothetical Turing Machine is a paradigmatic example of a classical computational system (Turing, 1997). In the present, we find a diversity of information processing systems. Cognitivist models are based on *classical computations*.

A classical computational system must first encode in symbolic tokens any nonsymbolic source of information (Piccinini & Shagrir, 2014). Essentially, this means that the system needs to transform an outside source (i.e., a physical event) that is meaningless for the system into something manageable by the syntactic rules of the system (Fodor, 1981). That is, the system needs to add semantic value to the physical input, thus creating an *input representation*. The computational system syntactically assigns the semantic value or meaning of an input representation (Fodor, 1975). Therefore, the system prespecifies the input value and is responsive only to those physical inputs it can put on its own terms. For cognitivism, these terms are symbolic and resemble the symbols of human language. The representations at work in cognitivism are hence *symbolic, language-like representations* (Haugeland, 1997).

In classical cognitivist models, computational systems transform input representations thanks to the application of sequences of instructions or algorithms that usually follow the binary rules of symbolic logic. After the information processing is complete, the system delivers a response that could be an *output representation*, that is, a representation useful for another module of the computational cognitive system, or a *motor command* to perform a bodily action, that is, a physical output. Susan Hurley (2001) called this basic design (input > information-processing > output) *the classical sandwich model of cognition* (Fig. 2.1).

Since, in the sandwich model, cognitive systems need the information sensory inputs provide and the bodily performance of its motor commands for interacting with the environment, we conceive cognitive processes as separated from any other environmental process. Hence, the mind becomes enclosed in the head's boundaries, as the brain is the machine responsible for the information processing.

Cognitivists not only appeal to computational models to explain the formal organization of cognition but also use these models to sketch brain processes at a macroscale. At the microscale, neuronal physicochemical activity is highly dynamic and complex because neurons exhibit different levels of hierarchical organization (Simon, 1962). Neuronal systems of the brain are usually interrelated and depend on each other to accomplish a cognitive function. However, for Simon, many neuronal systems are

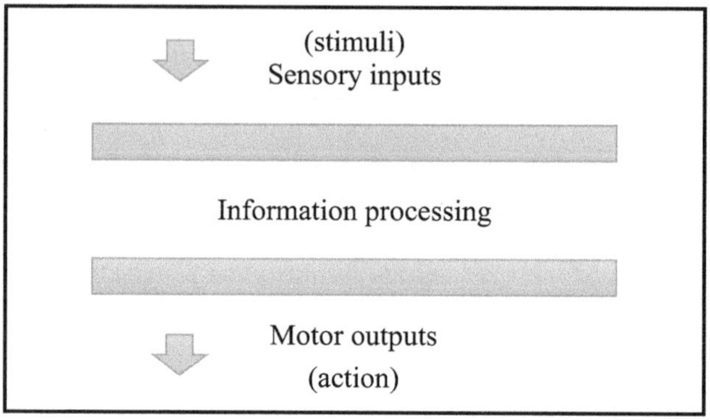

Fig. 2.1 A basic sketch of the classical sandwich model of cognition

decomposable. They are systems that we can treat as functional units that work relatively independently. At the macroscale, it means that brain processes exhibit specialized *functional modules* of information processing. These modules are interconnected linearly and form broader computational systems of information processing. From this perspective, the brain is a hyper-sophisticated computational system composed of multiple subspecialized and interconnected subsystems (Bassett & Gazzaniga, 2011). Consequently, the isomorphism of computational cognitive systems and brain processes interpreted computationally suggested that the computer-mind metaphor was a highly reliable scientific model of cognition.

2.1.2 Connectionism

Although cognitivist models seemed to describe brain processes accurately at the macroscale, it became increasingly evident to neuroscientists that things were different at the microscale. At this level, what matters is to explain the interactions between two neurons, or small clusters of neurons, instead of the functional modules described at the macroscale.

In the 1980s, a second major form of brain-centered cognitive science called *connectionism* addressed this problem by creating models of cognition based on interactive networks of artificial neurons. These models were valuable for creating solid hypotheses about the information processing at the microscale of brain activity. They also helped scientists to conceive cognition as a more dynamic, flexible, and complex phenomenon than assumed before.

The typical design of connectionist networks consists of three different layers of neurons: the input layer, the hidden layer, and the output layer. Input layers receive incoming stimuli, while the middle or hidden layer is responsible for information processing. Finally, the output layer delivers the response of the network (Fig. 2.2).

The environment provides stimuli to the input layers of the network. The hidden layer processes the information thanks to the emergence of patterns of activity on the web. These patterns are caused by the input and the already existent constraints of the network. The output layer delivers the system's response, produced by the emergent patterns of the web.

At the level of individual units, the activation of connectionist patterns follows Hebb's rule of neuronal activation. Hebb's rule indicates that if a neuron (A) receives an input from another neuron (B) at a time when both units are highly active, then the weight W(AB) to (A) from (B) should be strengthened (Rumelhart, 1997). Connections between units are thus strengthened thanks to mechanisms that allow the units to have some sort of memory and be more susceptible to activate in the presence of a one type of stimulus and not of others. This activity tries to emulate the physicochemical activation of biological neurons in the brain.

In the connectionist model, single units (isolated neurons) do not accomplish information processing. The collective activation of multiple neurons, producing a distributed activation pattern in the network, performs the information processing. The input layer triggers the web activity, but its existing connections constrain the pattern of activity. These

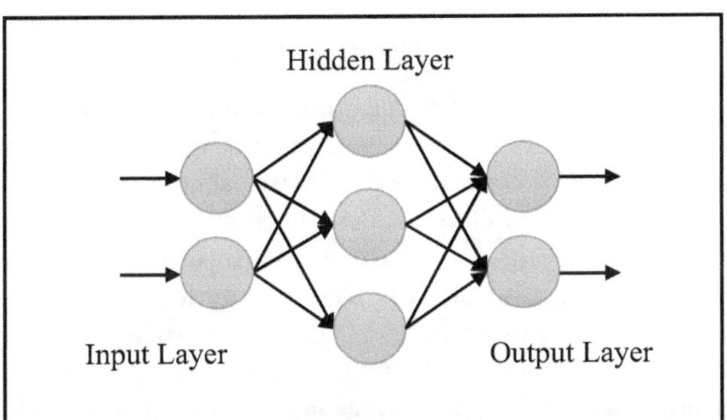

Fig. 2.2 A model of a basic connectionist network

connections depend partially on the network's physical design and the strength of the connections dynamically constructed between neurons. The regularity of the value of inputs will create regular patterns of activity across the web, delivering typical outputs. This activity is still open to dynamic changes if the input value constantly changes (Smolensky, 1997).

The open-endedness of these neuronal networks marks a crucial difference between cognitivist and connectionist models. Connectionist networks are based on rules of inference that handle statistical information instead of logical rules. Therefore, connectionist networks do not need to be preprogrammed like cognitivist models. They can handle dynamic changes in the environment and self-organize to adapt their responses to new circumstances, constantly establishing new patterns of activity. Connectionist networks are therefore more robust, flexible, and dynamic models than cognitivist ones.

Connectionist networks have a clear advantage over cognitivist models for explaining cognitive activities like learning, remembering, and recognizing patterns (Clark, 1997). These activities depend on more dynamic interactions with the environment, and connectionist networks do not require the extensive algorithmic programming of classical computations to process the information necessary for these activities. Connectionist models are, however, more imprecise than the cognitivist ones, especially for explaining more abstract cognitive functions like those at work in explicit reasoning and logical operations. For this reason, we standardly consider cognitivism the most successful scientific model of cognition thus far.

2.1.3 Predictive Processing

Predictive processing is the most contemporary form of brain-centered cognitive science. Like connectionism, predictive processing proposes a more dynamic model of cognition than classical cognitivism. Unlike connectionism, however, predictive processing does not conceive the brain as a passive system that waits for informational inputs to create patterns of neural activity. Instead, the brain is an active generator of these patterns that work as (statistical) hypotheses of the environment (Clark, 2013).

Brain hypotheses in predictive processing take the form of hierarchical statistical models called *generative models*. In this hierarchical order, the high-level models calculate the probabilities of responses from the low-level models. These latter models measure the probabilities of present and future sensorial inputs. Thus, the brain creates a statistical representation of the environment at different spatiotemporal scales (Wiese & Metzinger, 2017).

The acquisition of new information causes the constant adjustment of environmental models based on Bayes rules of inference. These Bayesian models recalculate prior probabilities considering new events or new information. The latest information creates a mismatch or *surprisal* between hypothesis and facts (Clark, 2013), triggering changes in the cascade of predictions from the higher to the lower-level models. The recalculation of probabilities has a high-level cost of energy (electrical energy in the case of our computers and metabolic energy in the case of the brain). Hence, instead of creating new models based on incoming information, the computational system only calculates the disparity between the expected and actual inputs. We call this process *prediction error minimization* (Hohwy, 2013; see also Fig. 2.3) and the compression of information (i.e., the focusing of the system on mismatches) *predictive coding* (Wiese & Metzinger, 2017).

Fig. 2.3 A very basic prediction error minimization model

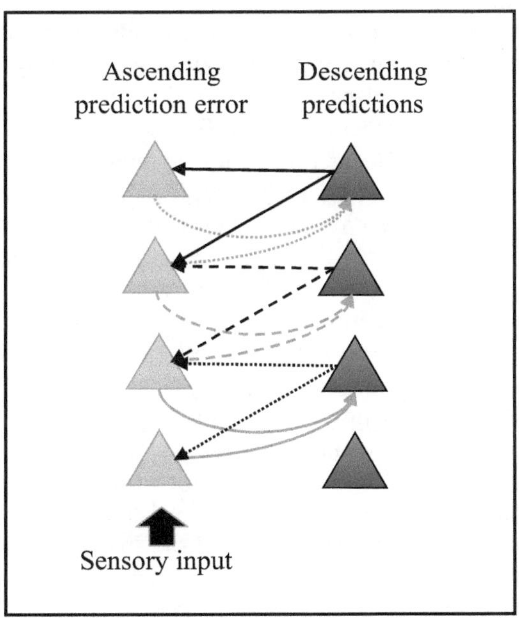

Black triangles represent the hierarchical generative models the brain creates as hypotheses of the environment in perceptual inference. Grey triangles represent the mismatch between the statistical expectation of the hypotheses and the current information provided by sensory input, that is, prediction error. The lines represent the feedback necessary between predictive models and surprisal to minimize error.

The flow of information between the brain, the body, and the world envisioned by predictive processing is so dynamic that some cognitive scientists suggest this is not a neurocentric model of cognition (Bruineberg et al., 2016; Clark, 2016; Friston et al., 2010). For this reason, we must distinguish between neurocentric and non-neurocentric forms of predictive processing (see Sects. 2.2.2 and 2.4.3). Here, I will focus on Hohwy's (2013, 2016) *prediction error minimization approach*, one of the predictive processing models strongly supporting neurocentrism.

Neurocentric interpretations of predictive processing are based on the belief that perception occurs in accord with Helmholtz's perceptual inference theory (Hohwy, 2013). Helmholtz argues that our brain makes inferences about the outside world based on the insufficient information our sensory organs provide. Conceived as an inferential machine-organ, the brain for Helmholtz is like a computational machine in the contemporary era. As in the sandwich model of cognitivism, the brain is isolated from the environment (Wiese & Metzinger, 2017). The only way to escape the black box is to maintain constant sensorimotor feedback with the environment (Clark, 2013).

Sensorimotor interaction in predictive processing nonetheless implies two different processes: *perceptual inference* and *active inference*. Perceptual inference describes the readjustment of brain hypotheses (generative models) that occur physically as changes in the constraints of neural connections (Clark, 2013). By contrast, active inference implies changes in the environment induced by the motor action of cognitive agents. Neurocentric predictive processing recognizes that bodily actions can change environmental conditions (either by changing the materiality of the environment or by changing the body's situation in the environment). Environmental changes cause alterations in the generative models of the brain (Hohwy, 2016). For this reason, sensorimotor loops, not just sensorial stimulation, causally constitute the statistical space of the brain models (Clark, 2016). However, neurocentric authors insist that cognition occurs in the brain exclusively because it is the brain and no other organ of the body that executes these information processing tasks (Hohwy, 2016). The body and the environment merely provide inputs and outputs. Therefore, predictive processing, like cognitivism and connectionism, reproduces the classical sandwich model of cognition.

2.2 The Prejudice of the Mind–World Dichotomy

Neurocentrism entails the assumption that the mind and the world are radically separated. Physically implemented in the human brain as a computational system, the mind is isolated from the extracranial surroundings. The brain can only communicate with the outside world thanks to the information provided by the senses and the motor commands it sends to the body. I have referred to this architectural description of cognition as the sandwich model.

Many empirical problems beset this model, and this section surveys some of them (Sect. 2.2.1). It also argues that the sandwich model of cognition is not intrinsic to all forms of computationalism and that neurocentrism is not a direct consequence of computationalism either. Therefore, we can separately analyze the pros and cons of neurocentrism and computationalism in cognitive science (Sect. 2.2.2). Since neurocentrism is not intrinsic to computationalism, I will also argue that the radical separation of mind and world is fundamentally based upon the theoretical prejudice of modern sciences of mind that we can call *the mind–world dichotomy* (Sect. 2.2.3). I will later show the emergence of empirical and philosophical problems because of this prejudice.

2.2.1 What Computers Could Not Do

The research program of cognitivism has probably been the most successful scientific approach to cognition in the history of cognitive science. Its computational models of mind have been helpful for scientific explanations in neuroscience, psychology, linguistics, artificial intelligence, and many other areas. Despite the success of cognitivism in too many areas, confining cognition to the boundaries of the head has turned out to be a very problematic assumption.

Cognitivist models' failures are usually due to isolating cognitive processes from the environment. In the field of artificial intelligence, for instance, the robotic agents of cognitivists have been incapable of sustaining efficient sensorimotor interactions with the environment. These failures have pushed cognitive scientists to look for alternatives to describe more dynamic interchanges between agents and the environment.

Under the principles of cognitivism, artificial intelligence models depend on producing symbolic representations and processing these representations with algorithmic rules. John Haugeland (1997) called these models *Good Old-Fashioned Artificial Intelligence* (GOFAI). GOFAI models successfully solved abstract operations like mathematical and logical problems. For example, computers playing chess at the level of the most skilled human players exhibited the potential of these types of artificial intelligence.

The success of GOFAI came to a halt when the limitations of the sandwich architecture of cognitivism became evident. Our everyday actions require the use of functions like perception and motor action. These functions do not seem to require abstract operations, at least not at the conscious level. Cognitivism hypothesizes, however, that computational processes are involved in these functions but at the sub-personal or unconscious level. Paradoxically, computational systems need to encode enormous amounts of information to perform the simplest tasks of action and perception. They also needed to be programmed with a vast number of instructions to process this information, much more than computers need to perform the abstract operations for playing chess (Dreyfus, 1972).

Theoretically, a computational system can handle everything that can be put into symbols and manipulated by algorithmic rules. Nevertheless, any real system in the physical world, even a computer, has spatiotemporal constraints. Thus, what may be, in principle, computable may not necessarily be computable given the actual spatiotemporal constraints of the physical world (Hendriks-Jansen, 1996). That is a problem for cognitivism because the amount of information processing its models needed to perform, even for the simplest functions of motor action and perception, was huge, and the programming is hardly feasible (Boden, 2016). As a result, GOFAI produced robots with abysmal performance, only capable of acting in highly controlled environments where the contingencies of the real-world environments were excluded.

Moreover, the fact that not only humans but other living organisms, even those with very simple or no nervous systems at all, *do* exhibit a great capacity to deal with the world in concrete bodily actions, without explicitly accomplishing lots of computations, put the hypotheses of the cognitivist program for explaining these aspects of cognition into question (Wheeler, 2005).

Furthermore, cognitivism faces at least two deeper sets of problems. First, living organisms exhibit the capacity to find objects of value (positive or negative) spontaneously and dynamically for their own purposes, such as when they recognize food, shelter, or potential threats. They seem to spontaneously experience some features of the world as more (or less) relevant than others. This is tantamount to saying that living organisms exhibit normative guided behavior (see Sect. 3.4). By contrast, cognitivist artificial intelligence robots need to be fully preprogrammed to attribute relevance to particular features of the world. Cognitivist robots cannot assign themselves value or meaning to the environment (Froese & Ziemke, 2009). This is a rough and straightforward description of what Harnad (1990) called *the symbol-grounding problem* in artificial intelligence.

Secondly, worldly situations are commonly complex and dynamic, and many of their components are inherently contingent. Hence, to recognize the relevant features of a real-world situation to accomplish a determinate task, a cognitive agent needs a great deal of cognitive agility and flexibility to recognize the situation and adapt its actions accordingly. However, robots constructed under cognitivist theories proved incapable of adapting to the contingencies of the environment. Consequently, they were unable to identify what might be relevant in a particular context. That is a rough description of the classical *frame problem* in artificial intelligence (Dennett, 1987).

The failures of GOFAI were not only problematic for creating efficient sensorimotor machines. They also exhibited the problems of cognitivism in explaining how cognition occurs in biological agents. For example, evolutionary psychology adopted cognitivist descriptions of the mind as a functional, modular system with preprogrammed responses (e.g., Cosmides & Tooby, 1987). For this type of psychology, living organisms' behavioral responses and psychological states, including humans, are determined by information encoded in genes and expressed in highly specialized brain modules (e.g., Pinker, 1998). This explanation, however, makes it challenging to account for the flexibility of organisms to adapt to different situations in their development. It is also hard to explain why organisms of the same species sometimes use very different strategies to solve the same problem (Lloyd, 1999). Indeed, the idea of a preprogrammed mind employing genetic information is no less problematic in biological organisms than in artificial systems (Hendriks-Jansen, 1996).

2.2.2 *Non-Neurocentric Computational Models*

Despite these practical problems of cognitivism, standard connectionism and prediction error minimization have insisted on locating cognition in the head. The main problem for these theories is not applying their alternative computational models to explain cognition formally but to interpret these models from a neurocentric stance.

Although connectionism is a more dynamic model than cognitivism, typical connectionists separate the cognitive domain from the outside world. Cognition happens in the network; the outside world merely provides inputs and receives outputs from the network. We may interpret the distributed pattern of activity in the network as distributed representations (Clark, 1997). Thus, the sandwich model is at play once again. Some defenders of cognitive neuroscience hold that we can produce hybrid computational models of cognition to overcome the discrepancies between the functional logic of cognitivist models at the macroscale of brain processes and the dynamical complexity of neuronal activity at the micro and mesoscales (P. S. Churchland & Sejnowski, 1988). The idea, in short, is that cognitivism and connectionism explain both mind and cognition in computational and neurocentric terms but at different explanatory levels (Harnad, 1990).

Connectionist information processing, however, conforms to the logic of dynamical systems theory (Smolensky, 1997). This theory has shown that non-neurocentric interpretations of cognition are possible (Port & van Gelder, 1995). Dynamical system theory models the change of a system that continuously evolves according to different variables. Sometimes these variables are other dynamical systems, and when two or more of these systems coevolve in time, constraining their behavior mutually, we say they are coupled. For coupling two dynamical systems, a system does not need to encode anything from the other; they only need to constrain their behavior (van Gelder, 1997). From this perspective, different neuronal networks can mutually constrain their activity (Freeman, 2000). Similarly, bodily motor actions and sensorial organs can also constrain the activity of neuronal networks (Kelso, 1995), while environmental processes can constrain agents' bodily actions and sensory feedback (Turvey & Carello, 1995). All this sustains the hypothesis that the brain, the body, and the environment form a dynamical system that works as a unitary whole.

Predictive processing also provides tools for modeling cognition as an embodied phenomenon. The free-energy principle is one of the pillars of predictive processing, and, according to Friston (2010), it describes processes of equilibrium between the interactional space of a living agent and the environment system rather than the creation of representational states in the brain. This interpretation undermines the neurocentric view of predictive processing.

The free-energy principle is a mathematical theory describing organisms as self-organizing systems that maintain low entropy levels, thus avoiding systemic disintegration while interacting with the environment to minimize their free energy (Friston, 2010). Entropy, in thermodynamics, is the measure of chaos and disorder of a system, whereas, from a statistical view, chaos is related to the unpredictability of states of a system. Entropy is therefore related to the informational measures of uncertainty and surprisal. Variational free energy measures the probabilities of a predictive or generative model to mismatch with the current sources of information, that is, of surprisal (see Sect. 2.1.3). Free energy is therefore equal to the amount of surprisal or uncertainty of a predictive model. Living organisms tend to reduce this informational uncertainty that is traduced physically in conserving low entropy levels (i.e., organisms stay alive).

One free-energy principle application is the brain's predictive processing model and cognition. As an organ that accomplishes free-energy reduction, the brain produces generative models that maintain low levels of uncertainty and entropy. However, we can also apply the free-energy principle to the interactions of the whole living organism with the environment (Allen & Friston, 2018). In this case, the generative model does not take the form of an internal representation that is physically implemented in the brain's neural structures. Instead, we can see that the whole organization of the organism, or its phenotype, is shaped by the processes of free-energy minimization. For this reason, supporters of the free-energy principle hold that the organism becomes the model of its environment or ecological niche (Ramstead et al., 2019).

In sum, connectionism and predictive processing involve computational models suitable from a non-neurocentric perspective (cf. Table 2.1). That means our computational models of cognitive processes do not force us to accept uncritically the claim that cognition is confined to the boundaries of the head. Neurocentrism is indeed a belief that some philosophers and scientists have endorsed even before the apparition of computational models. In the following subsection, I will argue that neurocentrism

Table 2.1 The primary forms of brain-centered cognitive science

	Computations	*Neurocentric interpretation*	*Non-neurocentric interpretation*
Cognitivism	Classical computation (symbolic & syntactic)	Encoding of information in symbols	N/A
Connectionism	Differential equations & dynamical systems theory	Neural information processing	Dynamical systems coupling (brain-body-environment)
Prediction error minimization	Bayesian statistical models	Brain segregation	Active inference

results from a prejudice of early scientists of mind inherited from the theoretical claims of modern philosophers. I will call this prejudice the mind–world dichotomy.

Table 2.1 summarizes the three main types of brain-centered cognitive science that have been discussed thus far, the type of computations of their models of cognition, and the neurocentric and non-neurocentric interpretations of these types of computations.

2.2.3 *The Mind–World Dichotomy*

The ***mind–world dichotomy*** is the assumption that the mind and the world are two object-like entities with radically different properties, they are independent of each other, and any relationship between the two is fundamentally causal and linear. The origin of the mind–world dichotomy is found in Descartes's ontological dualism, in which he describes mind and world as ontologically different entities (Rowlands, 2010).

Cartesian dualism, however, takes the form of a mind–world dichotomy once the gap between mind and world is adapted to the physicalist and reductionist frameworks of modern sciences of mind. ***Physicalism*** believes that the only things that exist are those entities and processes described by the science of physics (Neurath, 2012). A closely related position, ***reductionism***, implies that scientific laws are ultimately deducible from the laws of physics. That means the knowledge of all other sciences (such as biology, psychology, sociology, etc.) is, in principle, reducible to physics (E. Nagel, 1961). Although physicalism is an ontological hypothesis and reductionism an epistemological claim, many scientists and philosophers combine both claims into an *ontological reductionism* (Sarkar, 1992). This reductionism claims that the properties of all physical systems, from genomes to neutrinos, depending on the physical properties of the parts that compose these systems. The sciences of mind frequently assume physicalism and reductionism. Consequently, mental phenomena, just as any other phenomena, are seen as being, in principle, reducible to their underlying physical causes (Anderson, 2007).

There is, however, a gap between mind and world in at least two different senses. On the one hand, there is a distance between the mental realm circumscribed to the boundaries of the head, separated from the rest of the physical world (Rowlands, 2010). On the other hand, there is an explanatory gap between the causal explanations of mental phenomena and the experiential aspects of these phenomena that we can study from a subjective perspective alone (Chalmers, 1995). Since philosophers and scientists often conflate the two, the mind–world dichotomy causes multiple problems explaining mind and cognition. I briefly survey some of these problems in Sect. 2.3.

2.3 The Philosophical Problems of Neurocentrism

The empirical troubles of the sandwich model moved philosophers of cognitive science to question the philosophical bases of neurocentrism, thereby forcing a reappraisal of classical problems in philosophy of mind, such as representationalism (Sect. 2.3.1) and the explanatory gap (Sect. 2.3.2). Our motivations to abandon neurocentrism are thus not only scientific but also philosophical. In addition to the empirical and philosophical problems of neurocentrism, there are problems inherent to the methodologies of brain-centered cognitive science: These methodologies study cognition in the highly controlled environments of the lab (Sect. 2.3.3). Although these methodologies help identify some underlying causes of cognition, they miss significant aspects of this phenomenon. Many authors of embodied cognition have suggested incorporating methods that observe cognition as it happens in non-controlled environments. This strategy reveals that bodily and environmental processes are causally relevant for the constitution of cognition and not only what happens in the brain (Sect. 2.4).

2.3.1 *Representationalism*

Representationalism is probably the most frequently criticized issue of the neurocentric concepts of the mind. It originates in Franz Brentano's (Brentano, 1995) account of intentionality. As is well known, Brentano borrows the term "intentionality" from medieval philosophy. The word derives from the Latin *intentio,* which means "directed at." Intentionality describes the particular characteristic of mental states that, in contradistinction to physical phenomena, refer to something, and that is something other than themselves. In the philosophical jargon, intentionality concerns the *aboutness* of experiences: my perception, for instance, is *about* the maple tree in the garden of my neighbor, my memory is *of* my friend who has recently died, my imagining is *of* the Minotaur of Crete, and my thoughts are *about* who will win the presidential election in the United States. In these examples, we find that the objects my mental acts refer to have different modes of presentation. In some cases, the objects are currently present (e.g., the tree I see through the window). In other cases, they are absent (e.g., my dead friend), while in others, they are simply unreal (e.g., the Minotaur). An important lesson follows from this: In Brentano's eyes, intentionality does not imply the intended object's presence (or absence).

A closer analysis of the differences between these modes of presentation may be helpful to specify Brentano's position even more, for the contrast between the presence and absence of the intended object may not be as obvious as it first may seem. On the one hand, for non-present and unreal objects, the references of our mental acts cannot be physical objects that are currently present. So far, this is clear. On the other hand, although objects of perception may be physically present, their presence is only partial. The maple tree I see through the window, for instance, shows me only one of its visual profiles (e.g., the upper part of the tree), yet my perceptual experience is not about this piece of the tree but the whole tree. Therefore, it seems that my experiences *never* refer directly to physical objects, *not even in perception*; instead, they seem to be mediated by something *standing* between the physical reality of the world and subjective states of the mind. This is the birthplace of Brentano's representationalism.

For representational theories of mind, since mental acts refer to objects that are not present or not fully present, as in the case of perception, it follows that the mind cannot be directly related to material things of the world but to intermediary (intentional) objects that "represent" these objects (e.g., Searle, 1983). The contents of my experiences are, according to these theories, *copies* of physical reality; they are mental representations.[1]

That is clear, for representationalists, in the cases of imagination and memory, where the objects of my experience are not "in the flesh"—any experience of these objects *must*, they would argue, be representations. However, they further argue that the same is true in cases of perception, where my vision of the maple tree is not really about the physical tree but a mental representation of the tree. For this reason, even if the physical nature of my senses gives me only one profile of the object, my experience is about the whole object.

Many philosophers have suggested this and like hypotheses in the analytic tradition. They further support their representational accounts, appealing to Fregean semantics (Crane, 2001). Nonetheless, naturalized explanations of representationalism designate the connection between mental representations and physical objects causally (Fodor, 1981). In the same line of thought, philosophers like Ruth Millikan (1984) attribute the natural origin of this causal connection to processes of natural selection. Roughly speaking, these proposals suggest that a living organism possesses the capacity to represent the external physical world internally. Thanks to adaptation processes in its evolutionary past, the ensemble of its sensory organs and nervous system has evolved to accomplish these functional tasks.

A problematic version of Cartesianism exists in these representational theories of mind. Mental representations are usually described as object-like entities that possess an unclear ontological status that seems to differ from physical objects as such (Dreyfus, 1972). Natural processes might cause these representations, but it is hard to see if they also possess causal powers. The most common way of defining the role of mental representations in scientific explanations of mental phenomena is by reducing them to their functional role (Dretske, 1986). Representations have the function of carrying information; that is, they encode external physical features into symbolic *tokens* that the computational architecture of the mind can manipulate (Fodor, 1981). This definition, however, is not unproblematic, for it is based on the idea that the mind is a computational system that operates as software implemented by the hardware of the brain. Unfortunately, this division between software and hardware again imposes a dichotomy between physical and mental realms, just as Descartes did in his metaphysics.

There are also epistemological problems inherent to the mind–world dichotomy. Suppose the physicochemical activity of the brain is the direct cause of mental states. In that case, it is possible to imagine scenarios where the brain could be removed from its usual contact with the world and still have world-like experiences. Hilary Putnam envisioned the famous mental experiment of brains in a vat, a scenario where an evil scientist isolated our brains from our bodies and the rest of the world (Putnam, 1981). In this scenario, our brains can survive thanks to the nutrients provided by an artificial machine. Our experiences, by contrast, are caused by the physical stimuli produced by the same machine that emulates the stimulation of our sensory organs in normal conditions. It seems impossible to know with certainty what we are living, in what amounts to no more than a mere illusion in such a scenario.

The analogy is not simply whimsical. Whether subjective experiences are caused by the world or by an evil scientist's machine is irrelevant because we already live in an illusion of the world and an illusion of ourselves. For instance, for Metzinger (2003), all our conscious experiences of the world are simulations created by the brain that hardly correspond to the world as such.

From this standpoint, the world is nothing but physicochemical processes governed by physical laws. Our experiences are akin to hallucinations (Metzinger, 2003), mere by-products of brain processes (Anderson, 2007). Contrary to Descartes, not even the ego is real because our identities are also part of the hallucinatory realm of our minds (Metzinger, 2009). Subjective experiences such as perceptions, thoughts, emotions, and desires are ultimately irrelevant to the causal framework of the natural world.

The claim that the empirical world is a big illusion also puts the epistemological bases of all scientific knowledge in danger. Putting aside the skeptical arguments of foundationalist epistemologies (Carnap, 1969; Descartes, 2008), it is hard to understand how we can rely on our observations to support our scientific claims if we hold that the world we know is illusory. We first experience all the phenomena we make theories of. Even when objects like black holes are predicted by theoretical constructs before we directly or indirectly observe them, these phenomena need to make sense first in our conceptual schemes. That implies experiential aspects, whether they are mere numbers or abstract signs.

In the case of mental phenomena, things get worse. We do not begin our study of perception by looking into the retina or the brain's V1 zone. Instead, we start our scientific inquiries by asking how we perceive the colors red or green, or why my vision shows me only one object if I have two eyes, and so on. I can hypothesize that I can begin my scientific inquiries of cognitive phenomena in experience, and, once I got enough theoretical knowledge to elaborate my theories in purely causal terms, I can throw away my experiences, as eliminativists aim to do (P. Churchland, 1986). However, what is the guarantee that the experience where I began my scientific theories was accurate? If this experience is illusory, is it not a contradiction to use it as the origin of my explanations?

The problem with Metzinger's position is not obvious since we can claim that experiences are illusory *only* because the brain creates them and not because they are false. Experiences can connect us indirectly with the real world of raw physical processes. For instance, when I see the color red in a tomato, even if the tomato is not red, this experience still helps me study the perception of red things. I can trust my *false experience* of the tomato insofar as such an experience is a reliable basis for studying the physiological processes underlying this phenomenon. Still, the problem is that we can see differences in how different subjects experience the same event in their everyday lives and also in the history of philosophy and science (e.g., Mach's static description of a visual scene versus Gibson's dynamic description, cf. Noë, 2004). There is also empirical evidence that some aspects of perception are culturally relative (Pettersson, 1982). We must thus select which illusions are best for starting our scientific endeavors. Nevertheless, what then should be the criteria?

If we study subjective experience employing deductive methods, like Descartes, we risk remaining trapped in the world of illusions. The result is that it appears to be senseless for subjectivity to analyze itself. Maybe, as Ray Jackendoff (Jackendoff, 1987) proposes for cognitive science, the subjective experience should simply constrain all scientific labors and should be discarded once we have developed a well-supported hypothesis. Be this as it may, it still looks untenable to say that a world of illusions can guide knowledge that presumes to be the light at the exit of Plato's cave.

2.3.2 *The Explanatory Gap*

Neglecting subjective experience leads us to the second dimension of the mind–world dichotomy: the explanatory gap. In a physical world described by physicalism and reductionism, subjective experience is more an obstacle than a resource to studying mind and cognition. Contrary to our most immediate intuitions, our experience does not interact causally with the physical world for the mainstream sciences of mind. I may think, for instance, that my feeling of hunger makes me want to eat a strawberry. From the neurocentric perspective, however, my feelings cause nothing. My brain states result from information obtained from my body, and my response to this information causes me to look for this fruit and eat it. There is no need to invoke my "subjective feelings" in this story. Libet (1999) did experiments where he apparently "proved" that our brains make decisions just moments before we are consciously aware of them, thus reinforcing the view that subjectivity is superfluous.

This view fits all too well with reductionism. Indeed, since subjective experience is a non-observable phenomenon, and given that subjectivity is irrelevant to causal explanations of the mind, subjective experience is seen as a by-product of brain processes. With no causal interaction with the physical world, subjective experience becomes an *epiphenomenon*, that is, a phenomenon that merely supervenes on brain processes.

Epiphenomenalism and the correlated view of the world as a big illusion leave open some unsolvable problems. For example, even if, in principle, we can scientifically explain all the mechanisms that cause the emergence of mental states, we still cannot explain the particular feelings of these states, for example, how it feels like to undergo these states and no others. Jackson (1982) illustrates this problem in a hypothetical case of a scientist named Mary, who investigates and knows everything about the causal origins of the color red and our capacity to perceive this color. Nevertheless, Mary cannot know what it is like to experience the color red because she has lived in an environment where red is absent. This scenario puts one crucial aspect of mental phenomena (subjective experience) out of the reach of science and points to critical methodological aspects for studying the mind.

If the first-person perspective is inaccessible from a scientific standpoint, how can I know that other beings are sentient like me? This is commonly known as the problem of *other minds* (Dretske, 1973). To solve this problem, I can use inferences and simulations to explain that other beings that are similar to me likely possess the same kind of mental states and experiences that I do. This problem becomes more challenging, though, when we ask about the subjective feelings of other types of living organisms. This problem is exemplified by Nagel (1974), who wonders how it is possible to have epistemological access to the experience of an animal like a bat, which deploys very different sensorimotor interactions with the world.

Moreover, as Reber (2019) argues, this problem leads us to recent questions about the status of forms of artificial intelligence that display similar behavior and responses to human minds. The classical answer to the problem of mentality in artificial systems is the imitation game of Alan Turing (1997). He states that a machine can be seen as intelligent as long as the responses of an artificial device are similar enough to human reactions that we cannot discern whether this is a machine or a human being.

This solution has its critics. For instance, John Searle (1997) pointed to *the Chinese Room problem*. In this problem, Searle describes a scenario where a person is put in a box or a room and must translate sentences from English to Chinese. This person does not know any Chinese. To accomplish the task, she needs to use handouts and dictionaries to translate correctly from one language to the other without any idea of the meaning of the sentences she is translating. From the perspective of people outside the room, the person inside appears to know Chinese perfectly. The same happens when we attribute intelligence and mentality to machines that successfully perform tasks that we do, using our cognitive and mental capacities (such as *Google translate or Siri*, for instance). However, like in the case of the Chinese room, machines (or software like *Google translate or Siri*) do not *make sense* of what they are doing. The real problem is that we cannot affirm or deny mentality by appealing to our external observations alone; to do this, we need another criterion. All this seems to imply that accounting (even exhaustively) for causal explanations cannot suffice to explain mental phenomena.

2.3.3 Cognition in the Lab

As mentioned in Sect. 2.2.1, the early models of GOFAI were only capable of functioning within highly controlled environments. It was easier to construct a model to deal with a reduced number of variables than one that must dynamically change from one variable to another or deal with many variables at once. Real-world environments are full of contingencies, variables, shifts in practical contexts, and agents' motivations.

In this regard, the cognitivists Marvin Minsky and Seymour Papert (1988) recognized that artificial intelligence at the sensorimotor level could not use a general model of intelligence, as more classical cognitivists suggested (Newell & Simon, 1961). For Minsky and Papert, sensorimotor agents need to adapt their strategies to different environments and sets of variables in the same environment, according to the context of the sensorimotor task. Minsky and Papert called these different environmental contexts *microworlds*. For cognitivism, these conclusions did not make the task of cognitive science easier; instead, they exhibited the challenges of designing and programming efficient robotic agents. Controlling and simplifying environments is nonetheless a pervasive strategy of modern sciences of nature. Scientists must isolate phenomena to highlight the relevant causal variables they are trying to explain. The world in the lab is thus exceptional and artificial. Although this strategy is helpful for the analysis of certain phenomena, it seems to obscure the study of cognition.

Edwin Hutchins has been one of the most recalcitrant critics of this strategy of brain-centered cognitive science. As an anthropologist, Hutchins did not put the phenomenon he was studying into a controlled environment; instead, he got immersed in the field of cognition. His most remarkable work has been to analyze how multiple agents collectively accomplish the cognitive tasks required for the operation of a navy vessel (Hutchins, 1995). In this work, he described situations in which the vessel's procedures did not depend on the sum of the efforts of different cognitive agents but on dynamic interactions among multiple agents. These agents use various technological artifacts, producing distributed cognition patterns across the socio-material network that constitutes the vessel operation. The navigation of modern vessels, like many other phenomena, is so complex that we cannot recreate its conditions in the lab. Even if we try fragment phenomena and try to put the pieces under more controlled situations, reducing thus the number of variables to analyze, we risk losing sight of the nature of the phenomenon we are trying to explain.

Cognitivism also faced the problem of over-abstracting cognitive phenomena even in its area of most tremendous success: human reasoning. Initially, the models of human reasoning were highly abstract and normative (Newell et al., 1958). They tried to explain how human agents make decisions, solve problems in ideal conditions, and achieve their best performance. However, these models ignored real agents' limitations in obtaining all the information needed for optimal decision-making. They also neglect the constraints of time that push agents to elaborate strategies (heuristics) for taking actions without considering the information they already possess (Todd & Gigerenzer, 2003). Hence, the ideal scenarios of the early days of cognitivism needed to become more grounded in the real world to understand the concrete strategies of agents that deal with worldly situations. Simon described this new perspective as *bounded rationality* (Simon, 1972). More realistic models of rationality have progressively included other variables, for example, corporeal states, emotions, expectations, cultural habits, etc., limiting and shaping our rational capacities (Todd & Gigerenzer, 2003). All this shows that despite the undeniable progress made, the strategy of isolating cognitive phenomena in the lab, that is, over-abstracting the interactions of agents into highly controlled microworlds, has become gradually more and more problematic for cognitivism. The need for more dynamic models became more evident than ever.

2.4 Embodied Cognition: From Neurocentrism Revised to World-Involving Cognition

Hubert Dreyfus was one of the first philosophers to highlight the problems of cognitivist artificial intelligence or GOFAI. In his work, Dreyfus (1972) argued that cognitivists adopted mistaken biological, psychological, epistemological, and ontological assumptions, leading cognitive scientists to misunderstand the very nature of cognition. However, it was until the 1990s that Rodney A. Brooks revolutionized the engineering of mobile robots, implementing new architectures that addressed many of the problems that Dreyfus had diagnosed previously in GOFAI. Of distinctive significance is that in Brooks's (1991) designs, agents do not need to internalize the outside information but exploit the information already available in the environment. The emergence of Brooks' artificial intelligence was a crucial moment in the history of cognitive science. He proved that it was possible to model basic forms of cognition (sensorimotor cognition) in ways that reject the central presuppositions of neurocentric cognitive science. Indeed, his models were more efficient and simpler than those of cognitivism.

Varela, Thompson, and Rosch also published in 1991 *The Embodied Mind*. This groundbreaking work rejected the main philosophical presuppositions of brain-centered cognitive science and looked for a new way to conceive and study cognition. Around this time, Antonio Damasio (1994) questioned the Cartesian background of mainstream neuroscience that assumed the mind–world dichotomy and neglected the crucial role of emotions and affects on the activities of cognitive agents. In the same decade, models of cognition based on dynamical systems theory began to study perception as a dynamical process that coupled the brain, the body, and the environment in a single complex system (Kelso, 1995). Dynamical systems theory was also helpful for explaining the attunement of the body and the environment in sensorimotor interactions without the intervention of internal representations (Turvey & Carello, 1995). At the end of the 1990s, Andy Clark and David Chalmers (1998) took the idea of embodied cognition even further. These philosophers made the controversial claim that our minds can be spatially localized beyond the boundaries of the brain when we use artifacts to expand our biological and cognitive skills. Consequently, brain-centered cognitive science was no longer the only game in town at the dawn of the new millennium. However, the different theoretical backgrounds and agendas behind the various forms of embodied cognition have impeded the unification of a single research program as an alternative to brain-centered cognitive science (Shapiro, 2010).

There are many possible ways to categorize the different research programs on embodied cognition. Here, I find it relevant to draw a division considering the degree of opposition these programs show to the dogmas of brain-centered cognitive science. Therefore, like Hutto and Myin (2013) and Gallagher (2017), I distinguish three main types of embodied cognition that I will call *weak, moderate, and radical.* In what follows, I will survey these three alternatives (Sects. 2.4.1–2.4.3) and argue that only radical embodied cognition is a truly *world-involving cognition*. This branch is the only one that genuinely overcomes the mind–world dichotomy and treats cognition as a phenomenon deeply interconnected to the world that surrounds cognitive agents (Sect. 2.4.4).

2.4.1 Weak Embodied Cognition

Weak embodied cognition acknowledges that abstract forms of cognition like imagination and thought are grounded on more basic sensorimotor levels. However, they conceive the sensorimotor level of cognition as a particular set of brain processes, different from those supporting more abstract forms of cognition. Weak embodied cognition hence does not look at the concrete dynamic processes that cut across the brain, the body, and the environment to explain the emergence of cycles of action and perception. On the contrary, they appear to subscribe to the same distinction that the opposite view works with. A paradigmatic example of weak embodied cognition is the theory of *b-format representations.*

Goldman and Vignemont (2009) found that embodied cognition comprises a wide variety of approaches, not all equally viable for a scientific approach to cognition. While some approaches trivialized the claim that cognition is embodied, others had no clear theoretical base for supporting it. These authors, therefore, proposed a "sanitized" theory of embodied cognition that involved the existence of a particular type of representation, corresponding to sensorimotor areas and pathways of the brain that they called b-format representations. These representations have the peculiarity of being active either in action, perception, or in processes of empathy, with the motor actions of other agents (e.g., mirror neurons: Rizzolatti & Craighero, 2004). Moreover, the same representations of sensorimotor activate in higher cognitive activities such as thought and imagination if such activities are related to the contents of action and perception (Goldman, 2012).

Like standard definitions of internal representations, b-format representations theory entails the belief that, while accomplishing cognitive activities, our brains need to use information that is not available in the environment at the moment. This information is contained in representational formats and physically implemented in brain processes and structures. Therefore, the defenders of b-format representations adopt the same sandwich model of cognition described by classical representationalists. However, there are many reasons for challenging this theory that neglects the dynamicity and complexity of social and sensorimotor interactions (Gallagher, 2017).

Although interactions with other people can trigger activity in sensorimotor areas in the brain, this does not entail that the bodily movement of other people is internally represented. It just shows that bodily interactions are the basis of social encounters. First, it is not only the brain but also the whole body that gets prepared for action when we look at others realizing a bodily action (Frijda, 1986). Muscular tensions, the rhythm of heartbeats, glandular segregations, etc., are part of our action-readiness when we perceive other people's bodily actions, not only neural pathways (Bruineberg & Rietveld, 2014; Colombetti, 2014).

Second, we cannot internalize other people's responses in representational formats due to the highly unpredictable and dynamic changes in their actions (Gallagher, 2015). Interacting with other agents demands unforeseeable and creative answers from agents and emotional and expressive (gestural) aspects that challenge the possibility of decoupling an agent's actions, perceptions, emotions, and expressions from the same kind of activities produced by others (Gallagher, 2017).

Third, dynamicity and unpredictability are not exclusive to social interactions. Environmental situations also change constantly in cycles of actions and perception. For classical representational models of action and perception, cognition consists of interpreting perceptual data from sensorial inputs and sending motor commands to motor outputs so that agents can manage the contingencies of environmental situations (Pylyshyn, 1984). These explanations, however, neglect the environmental and bodily contingencies that produce continuous (online) variations and adjustments in sensorimotor loops (Clark, 1997). Perceptually, the flow of sensorial stimuli dynamically changes thanks to the body's action (Thelen & Smith, 1994). Bodily actions also vary dynamically due to multiple contextual variables of the environment (e.g., pressure and inertia) and of the body itself (e.g., energy levels and emotional states). Sensorimotor systems must constantly change and adapt dynamically to the circumstances, establishing a dynamic coupling with the environment. This coupling cannot be encapsulated in representational formats, at least not in the traditional manner (Gallagher, 2017).[2] These conclusions do not deny the significant role of the brain in sensorimotor and social interactions. However, they lead us to see that brain processes are just one fragment of the transcranial body–environment system. The brain actively modulates the interactional system in this system, but it does not need to internally represent the world or the body for this task (Fuchs, 2018).

The solution of weak embodied cognition thus looks more like the classical use of *ad hoc* strategies to maintain the theoretical core of a scientific theory. Weak embodied cognition changes nothing regarding the sandwich model of cognition. It pictures cognition again as the processing of encapsulated vehicles of information detached from the concrete body and, consequently, from the rest of the environment.

2.4.2 Moderate Embodied Cognition

Moderate embodied cognition, in contrast to weak embodied cognition, accepts that cognition is not restricted to the brain since there are many cases where cognition involves the coupling of the brain–body–environment system as a single functional unit. Nonetheless, this approach still endorses the claim that cognition consists of information processing (Wilson & Clark, 2009) and typically occurs in the human brain. Moderate embodied cognition thereby rejects neurocentrism but endorses computationalism. The hypothesis of the extended mind (henceforth "the extended mind") is one of the best examples of moderate embodied cognition in this regard.

The extended mind claims that cognitive systems are *occasionally* constituted by the brain, bodily, and environmental structures (Clark & Chalmers, 1998). The defenders of this hypothesis base their arguments on a functionalist perspective that they call *extended* or *embodied functionalism* (Wheeler, 2010). For cognitivism, computational cognitive functions are physically implemented in the brain exclusively. However, for proponents of extended functionalism, information processing is causally supported by extended systems that cut across the brain, body, and the environment. Paradigmatic examples of extended mind are using notes as memory cues or a piece of paper for mathematical operations. Extended functionalism is not committed to any particular form of computationalism but assumes that cognition involves some form of information processing. The revolutionary idea is that information processing can be physically implemented in very different systems, not exclusively in the human brain.

Extended functionalism nonetheless entails problems for both neurocentric cognitive science and radical embodied cognition. From a neurocentric perspective, cognition does not merely involve information processing functions but also the spontaneous generation of semantic representations needed to accomplish computational cognitive functions (Adams & Aizawa, 2008). No physical systems exhibit the spontaneous production of semantic representations other than the human brain (Adams & Aizawa, 2008; Searle, 1997). Therefore, from a neurocentric perspective, the bodily and environmental processes involved in cognitive tasks do not fulfill the criteria of cognitive functions. These processes are better seen as auxiliary causal processes of cognition (Rupert, 2009).

From the standpoint of radical embodied cognition, the extended mind is no less problematic since extended functionalists still endorse computationalism, which radical embodied cognition rejects. On its grounds, from this standpoint, cognition is rooted in the sensorimotor interactions of an agent and the environment. We can perfectly model these interactions in non-representational models of dynamical systems theory. These models do not need to assume a radical separation between the brain, the body, and the environment. Therefore, as Hutto and Myin (2013) claim, from a radical perspective, minds are not extended but *extensive* because their constitution implies the coupling of the brain, body, and environment as a unitary system from the start. Moreover, for some radicals, the living body of cognitive agents plays a role that breaks the symmetry with the environment, modulating their interactions with the environment (Thompson &

Stapleton, 2009). That means the structures and processes of the environment cannot play the same role as brain and bodily processes, which is one of the claims made by extended functionalism.

2.4.3 Radical Embodied Cognition

Radical embodied cognition explicitly rejects the theses of computationalism and neurocentrism. Although radical embodied cognition has principally focused on the rejection of representationalism and not computationalism, the rejection of the mind–word dichotomy led cognitive scientists to redefine the whole phenomenon of cognition. We find this to be the case in the two primary forms of radical embodied cognition: *enactive cognitive science* and *ecological approaches to cognition* (cf. Fig. 2.4).

Enactive cognition or enactive cognitive science originated in the claims of Varela, Thompson, and Rosch in *The Embodied Mind* (2016), where they argued that cognition was not information processing but *enaction*. This idea is more complex than many cognitive scientists usually assume and needs a deep examination to reveal its philosophical implications. I will take up this task in Sect. 3.1. In the meantime, we can say that

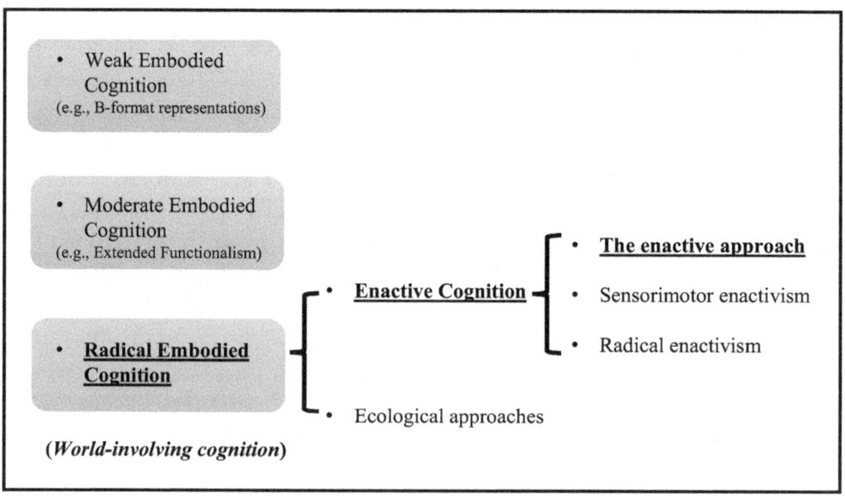

Fig. 2.4 The main branches of embodied cognition

enaction basically consists of a twofold claim: (1) perception is based on regular patterns of sensorimotor interaction between an agent and the environment, and (2) all forms of cognition are ultimately grounded in the same sensorimotor patterns. Defining cognition as enaction thus leads to rejecting the notion that perception is representational and computational because sensorimotor patterns are concrete "happenings" in the world that depend on the concrete bodily actions of agents. Dynamical systems theory (Thompson & Varela, 2001) and, more lately and arguably, the free-energy principle (Kirchhoff & Froese, 2017) have reinforced this claim on an empirical basis.

Nowadays, many proposals of radical embodied cognition claim to be enactive. The enactive approach is the first and foremost form of enactive cognition, which is most known as *autopoietic enactivism*. The central idea of this approach is that cognition is based on the same organizational principles of life, and this idea originated in the theory of autopoiesis. However, the theory of biological autonomy the supporters of the enactive approach currently defend has evolved and does not coincide with the original autopoiesis theory. The scholars of the enactive approach strongly reject this label (e.g., Barandiaran, 2017; Thompson, 2018), and, for this reason, I am not using it here. I will argue later this approach is the most reliable form of embodied cognition because it is the one that puts more radically in question the philosophical presuppositions of brain-centered cognitive science. *Sensorimotor enactivism* is, chronologically speaking, the second major branch of enactive cognitive science, mainly based on the theory of *sensorimotor contingencies* (O'Regan & Noë, 2001). This theory describes perception as acquiring practical knowledge or know-how of lawful sensorimotor correlations. *Radical enactivism,* the third primary form of enactive cognitive science, undermined representational theories of mind and cognition by demonstrating how an account of information without representational content, information as co-variation, helps us understand the sensorimotor basis of cognition (Hutto & Myin, 2013, 2017). Other philosophers and scientists, most notably Shaun Gallagher (2005, 2017) and Thomas Fuchs (2017, 2018), have also made significant contributions to modeling sensorimotor and social cognition from a non-computational and non-neurocentric point of view that we can fairly call enactive. Nevertheless, these authors have not been explicitly committed to the main principles of one of the three primary forms of enactivism.

Although the focus of enactive cognition was originally on how individual bodily actions causally constitute cognition, there is now an increasing interest in how social interactions and environmental structures are constitutive of this phenomenon (Gallagher, 2017; Hutto & Myin, 2017; McGann, 2014). This subject has been significant for enactive cognition to explain how cognitive activities, seen as properly human (e.g., imagination, symbolic language, and abstract reasoning), have scaled up from basic cycles of action and perception. The study of the role of environmental structures for cognition has, nonetheless, been the focus of the other main branch of radical embodied cognition, that of *ecological approaches*.

Ecological approaches to cognition originate in the work of the American psychologist James J. Gibson, who created a non-representational and non-neurocentric theory of visual perception almost simultaneously with the emergence of cognitive science. For Gibson, perceivers do not need to represent information internally because the environment already has an informational structure accessible to organisms through locomotion (Gibson, 2015). Gibson called this informational structure *ecological information,* and it is on this informational basis that perceivers can access the most primitive forms of perceptual meaning, which Gibson calls *affordances.* In this regard, our perception is primarily guided by what things afford as *possibilities for actions.* The chair affords sit-ability, this cup affords grasp-ability, and so on. These pragmatic "meanings of things" are affordances, and Gibson's thesis is that these are *perceptually* accessible to the agent.

Michael Turvey (1992), Edward Reed (1996), Harry Heft (2015), Anthony Chemero (2009), and other ecological psychologists have continued to develop the original concepts of Gibson to make ecological psychology a solid scientific theory that rivals mainstream forms of sciences of mind. The skilled intentionality framework is probably the most systematic and complete ecological approach to cognition (Rietveld et al., 2018). Adapting the framework of ecological psychology to address Wittgenstein's account of sociocultural practices, this approach has constructed a robust theory of skillful human practices. Contrary to other ecological approaches, this approach has incorporated a theory of how individual agents become attuned to ecological frameworks that are both natural and cultural. As a theory of bodily attunement, skilled intentionality works from an embodied interpretation of the free-energy principle (Friston, 2010) and the phenomenological work of Merleau-Ponty (2012). The notion of skilled intentionality strongly resonates with the general view of enactive cognition since it describes how agents enact a sensorimotor domain of interactions based on the natural and social constraints that shape their interactional space (Kiverstein & Rietveld, 2018). The complementarity between enactive and ecological approaches for a unified and more robust account of radical embodied cognition seems necessary. However, in the following chapters, I will consider significant challenges to this task.

Radical embodied cognition is not exempted from problems and criticisms. Since thus far enactive cognition has mainly focused on basic forms of cognition, some authors accuse enactivists of reviving behaviorism and explaining phenomena that we can hardly call cognition (Aizawa, 2014). There is also a challenge of scaling up explanations of the basic levels of cognition based on non-representational models to the abstract complex forms that seem to imply the use of internal representations (Clark & Toribio, 1994). Many of these problems are nonetheless proper for scientific programs that are still under construction. There is no reason to think that enactivist cannot address and surmount those issues soon.

2.4.4 *Cognition without Dichotomies*

Embodied cognition comprises various proposals that offer different alternatives to brain-centered cognitive science. We can evaluate the success of each kind of embodied cognition depending on the support they have from empirical evidence, their theoretical coherence, the number of predictions they can produce, the practical application of their ideas, etc. Here I evaluate them by their potential capacity to overcome the mind–world dichotomy that has been a highly problematic theoretical assumption for brain-centered cognitive science. I hold that closing the gap between mind and world can lead us to establish better scientific and philosophical problems to explain and understand the nature of the embodied mind.

As we have seen, not all forms of embodied cognition reject the two central dogmas of brain-centered cognitive science: neurocentrism and computationalism. Weak embodied cognition still locates cognition in the head and insists on using internal representations to explain all forms of cognition, including sensorimotor functions. Despite the subtle inclusion of sensorimotor processes as the foundations of cognition, the gap between the mind and the world seems to be the same that brain-centered cognitive science assumes.

Moderate embodied cognition moves away from the neurocentrism of brain-centered cognitive science, describing cognitive processes that can happen in extended systems that cut across the brain, the body, and the environment. However, accepting that cognition consists of information processing has some problematic implications.

On the one hand, the first versions of the extended mind not only imply the definition of cognition as information processing, that is, endorsing computationalism, it also assumed that the core of cognition happens as brain-centered cognitive science told us. The innovative idea of the extended mind was only that external resources could improve (neurocentric) cognition by offering complementary and additional forms of information processing and informational storage. That does not overcome the fundamental mind–world dichotomy because the cognitive agent is still subject to isolation in scenarios like the brain-in-a-vat experiment. In such a view, nothing prevents us from the possibility of being a brain in a vat. We might be isolated organs performing computational but interconnected to a more extensive computational system, like when we branch our personal computers to the World Wide Web.

The explanatory gap also remains intact for any approach that endorses computationalism. If cognition consists of information processing, then any machine that accomplishes this function in the way humans seem to do it can be considered cognitive and intelligent. Nowadays, artificial intelligence systems exhibit an extraordinary capacity to interact linguistically with human agents and can easily fool any of us, passing the Turing test. That does not mean they are intelligent or cognitive since we cannot exclude the possibility of the Chinese room problem. Nothing can tell us in these functional computational terms that machines are genuinely sentient. It is not enough to be fooled by their capacity to articulate coherent and unexpected responses that seem to exhibit real intelligence.

Indeed, the enactive approach is the only form of embodied cognition that truly offers a definition of cognition that can overcome these problems. The enactive approach defines cognition as sense-making, and this activity is the unavoidable consequence of what this approach defines as the autonomy of life. This theory found that cognition is not only rooted in a sensorimotor basis but also the organizational nature of the living body. This type of body is by definition sentient and to be sentient and alive seems to be the root of any form of real intelligence. There is no doubt that these ideas are controversial and still need further support from theoretical and evolutionary biology. However, these provocative ideas have already shown is potential to understand cognition from a wider stance than brain-centered cognitive science. Next to other approaches to world-involving cognition, the enactive approach can lead us to a revolutionary view of cognitive science that overcomes the problematic mind–world dichotomy.

In Chap. 3, we'll examine the core ideas of enactive cognition and why the enactive approach is, in my view, the most promising form of world-involving cognition.

Notes

1. This sketch of representational theories of mind may caricaturize the proposals of these theories. However, since there is a wide corpus of representational theories that holds different notions of representations and different manners to argue in favor of a mediated relation between mind and world, a deeper description of these ideas would implicate an extensive analysis that goes beyond the scope and interest of this book. It is not my aim to reject representationalism but to show the problems this theoretical standpoint has carried for cognitive science, motivating thus the emergence of embodied cognition. By contrast, I think that the general formulation I'm offering apply to the most standard theories of mental representations
2. In moderate versions of embodied cognition, there are also hypotheses that consider the existence of special types of representations (minimal representations) like predictive emulators (Clark & Grush, 1999) or action-oriented representations (Wheeler, 2005). Although this kind of hypotheses are sound because they highlight that we always need more than real-time or online dynamics to explain the efficient coordination of cycles of action and perception, they also entail a much broader conception of representations than the classical ones. In these versions, representations are either highly dynamic and plastic to be seen as vehicles decouplable from the brain–body–environment system (Thompson, 2007), or they are conceived as vehicles implemented all across this system (Wheeler, 2005, p.193). If this is the case, Gallagher (2017) argues it might be better to abandon the talk of representations because is useless and recognize instead the existence of structural dynamic wholes (*Gestalten*) that predispose the brain–body–environment system to acquire a particular dynamic coordination (Gallagher et al., 2013).

References

Adams, F., & Aizawa, K. (2008). *The bounds of cognition*. Blackwell Publishing.

Aizawa, K. (2014). The enactivist revolution. *AVANT, 5*(2), 19–42.

Allen, M., & Friston, K. J. (2018). From cognitivism to autopoiesis: Towards a computational framework for the embodied mind. *Synthese, 195*(6), 2459–2482.

Anderson, J. R. (2007). How can the human mind occur? In *How can the human mind occur in the physical universe?* (pp. 237–248). : Oxford University Press.
Barandiaran, X. E. (2017). Autonomy and enactivism: Towards a theory of sensorimotor autonomous agency. *Topoi, 36*(3), 409–430.
Bassett, D. S., & Gazzaniga, M. S. (2011). Understanding complexity in the human brain. *Trends in Cognitive Sciences, 15*(5), 200–209.
Boden, M. A. (2016). *AI: Its nature and future.* Oxford University Press.
Brentano, F. (1995). *Psychology from an empirical standpoint.* (A. C. Rancurello, D. B. Terrell, & L. L. McAlister, Trans.). New York, NY: Routledge.
Brooks, R. A. (1991). Intelligence without representation. *Artificial Intelligence, 47*(1–3), 139–159.
Bruineberg, J., Kiverstein, J., & Rietveld, E. (2016). The anticipating brain is not a scientist: The free-energy principle from an ecological-enactive perspective. *Synthese, 195*(6), 2417–2444.
Bruineberg, J., & Rietveld, E. (2014). Self-organization, free energy minimization, and optimal grip on a field of affordances. *Frontiers in Human Neuroscience, 8*, 599. https://doi.org/10.3389/fnhum.2014.00599
Carnap, R. (1969). *The logical structure of the world & pseudoproblems in philosophy.* (R. A. George, Trans.). Berkeley: University of California Press.
Chalmers, D. J. (1995). Facing up to the problem of consciousness. *Journal of Consciousness Studies, 2*(3), 200–219.
Chemero, A. (2009). *Radical embodied cognitive science.* MIT Press.
Churchland, P. (1986). *Scientific realism and the plasticity of mind.* Cambridge University Press.
Churchland, P. S., & Sejnowski, T. J. (1988). Perspectives on cognitive neuroscience. *Science, 242*(4879), 741–745.
Clark, A. (1997). *Being there: Putting brain, body, and world together again.* MIT Press.
Clark, A. (2013). Whatever next? Predictive brains, situated agents, and the future of cognitive science. *Behavioral and Brain Sciences, 36*(3), 181–204.
Clark, A. (2016). *Surfing uncertainty: Prediction, action, and the embodied mind.* Oxford University Press.
Clark, A., & Chalmers, D. J. (1998). The extended mind. *Analysis, 58*(1), 7–19.
Clark, A., & Grush, R. (1999). Towards a cognitive robotics. *Adaptive Behavior, 7*(1), 5–16.
Clark, A., & Toribio, J. (1994). Doing without representing? *Synthese, 101*(3), 401–431.
Colombetti, G. (2014). *The feeling body: Affective science meets the enactive mind.* The MIT Press.
Cosmides, L., & Tooby, J. (1987). From evolution to behavior: Evolutionary psychology as the missing link. In J. Dupré (Ed.), *The latest on the best: Essays on evolution and optimality* (pp. 276–306). The MIT Press.

Crane, T. (2001). *Elements of mind: An introduction to the philosophy of mind.* Oxford University Press.

Damasio, A. R. (1994). *Descartes' error: Emotion, reason, and the human brain.* G.P. Putnam's Sons.

Dennett, D. C. (1987). Cognitive wheels: The frame problem of AI. In Z. W. Pylyshyn (Ed.), *The Robot's Dilemma revisited: The frame problem in artificial intelligence* (pp. 41–64). Ablex Publishing Corporation.

Descartes, R. (2008). *Meditations on first philosophy: With selections from the objections and replies.* (M. Moriarty, Trans.). Oxford: Oxford University Press.

Dretske, F. (1973). Perception and other minds. *Noûs, 7*(1), 34–44.

Dretske, F. (1986). Misrepresentation. In R. J. Bogdan (Ed.), *Belief: Form, content, and function* (pp. 17–36). Clarendon Press.

Dreyfus, H. L. (1972). *What computers can't do: A critique of artificial reason.* Harper & Raw Publishers.

Fodor, J. A. (1975). *The language of thought.* Thomas Y Crowell Company.

Fodor, J. A. (1981). *Representations: Philosophical essays on the foundations of cognitive science.* MIT Press.

Freeman, W. J. (2000). *How brains make up their minds.* Columbia University Press.

Frijda, N. H. (1986). *The emotions.* Cambridge University Press.

Friston, K. J. (2010). The free-energy principle: A unified brain theory? *Nature Reviews Neuroscience, 11*(2), 127–138.

Friston, K. J., Daunizeau, J., Kilner, J., & Kiebel, S. J. (2010). Action and behavior: A free-energy formulation. *Biological Cybernetics, 102*(3), 227–260.

Froese, T., & Ziemke, T. (2009). Enactive artificial intelligence: Investigating the systemic organization of life and mind. *Artificial Intelligence, 173*(3), 466–500. https://doi.org/10.1016/j.artint.2008.12.001

Fuchs, T. (2017). Intercorporeality and Interaffectivity. In C. Meyer, J. Streeck, & J. S. Jordan (Eds.), *Intercorporeality: Emerging socialities in interaction.* Oxford University Press.

Fuchs, T. (2018). *Ecology of the brain: The phenomenology and biology of the embodied mind.* Oxford University Press.

Gallagher, S. (2005). *How the body shapes the mind.* Oxford University Press.

Gallagher, S. (2015). Invasion of the body snatchers: How embodied cognition is being disembodied. *The Philosophers' Magazine, 68*, 96–102.

Gallagher, S. (2017). *Enactivist interventions: Rethinking the mind.* Oxford University Press.

Gallagher, S., Hutto, D. D., Slaby, J., & Cole, J. (2013). The brain as part of an enactive system. *Behavioral and Brain Sciences, 36*(4), 421–422. https://doi.org/10.1017/S0140525X12002105

Gibson, J. J. (2015). *The ecological approach to visual perception.* Psychology Press.

Goldman, A. (2012). A moderate approach to embodied cognitive science. *Review of Philosophy and Psychology, 3*(1), 71–88.

Goldman, A., & de Vignemont, F. (2009). Is social cognition embodied? *Trends in Cognitive Sciences, 13*(4), 154–159.
Harnad, S. (1990). The symbol grounding problem. *Physica D: Nonlinear Phenomena, 42*(1–3), 335–346.
Haugeland, J. (1997). What is mind design? In J. Haugeland (Ed.), *Mind design II: Philosophy, psychology, artificial intelligence* (pp. 1–28). MIT Press.
Heft, H. (2015). *Ecological psychology in context: James Gibson, Roger Barker, and the Legacy of William James's Radical Empiricism*. Psychology Press.
Hendriks-Jansen, H. (1996). *Catching ourselves in the act: Situated activity, interactive emergence, evolution, and human thought*. MIT Press.
Hohwy, J. (2013). *The predictive mind*. Oxford University Press.
Hohwy, J. (2016). The self-evidencing brain. *Noûs, 50*(2), 259–285.
Hurley, S. (2001). Perception and action: Alternative views. *Synthese, 129*(1), 3–40.
Hutchins, E. (1995). *Cognition in the wild*. MIT Press.
Hutto, D. D., & Myin, E. (2013). *Radicalizing enactivism: Basic minds without content*. MIT Press.
Hutto, D. D., & Myin, E. (2017). *Evolving enactivism: Basic minds meet content*. MIT press.
Jackendoff, R. (1987). *Consciousness and the computational mind*. MIT Press.
Jackson, F. (1982). Epiphenomenal qualia. *The Philosophical Quarterly, 32*, 127–136.
Kelso, J. A. S. (1995). *Dynamic patterns: The self-organization of brain and behavior*. MIT Press.
Kirchhoff, M. D., & Froese, T. (2017). Where there is life there is mind: In support of a strong life-mind continuity thesis. *Entropy, 19*(4), 169.
Kiverstein, J. D., & Rietveld, E. (2018). Reconceiving representation-hungry cognition: An ecological-enactive proposal. *Adaptive Behavior, 26*(4), 147–163.
Libet, B. (1999). Do we have free will? *Journal of Consciousness Studies, 6*(8–9), 47–57.
Lloyd, E. A. (1999). Evolutionary psychology: The burdens of proof. *Biology and Philosophy, 14*(2), 211–233.
McCulloch, W. S. (1949). The brain computing machine. *Electrical Engineering, 68*(6), 492–497.
McGann, M. (2014). Enacting a social ecology: Radically embodied intersubjectivity. *Frontiers in Psychology, 5*, 1321.
Merleau-Ponty, M. (2012). *Phenomenology of perception*. (D. Landes, Trans.). New York, NY: Routledge.
Metzinger, T. (2003). *Being no one: The self-model theory of subjectivity*. MIT Press.
Metzinger, T. (2009). *The ego tunnel: The science of the mind and the myth of the self*. Basic Books.
Millikan, R. G. (1984). *Language, thought, and other biological categories*. MIT Press.

Minsky, M., & Papert, S. A. (1988). *Perceptrons*. MIT Press.
Nagel, E. (1961). *The structure of science: Problems in the logic of scientific explanation*. Harcourt, Brace & World.
Nagel, T. (1974). What is it like to be a bat? *The Philosophical Review, 83*(4), 435–450.
Neurath, O. (2012). *Philosophical papers 1913–1946: With a bibliography of Neurath in English*. (R. S. Cohen & M. Neurath, Trans.). Dordrecht, NL: D. Reidel Publishing Company.
Newell, A., Shaw, J. C., & Simon, H. A. (1958). Elements of a theory of human problem solving. *Psychological Review, 65*(3), 151.
Newell, A., & Simon, H. A. (1961). Computer simulation of human thinking. *Science, 134*(3495), 2011–2017.
Noë, A. (2004). *Action in perception*. MIT Press.
O'Regan, J. K., & Noë, A. (2001). A sensorimotor account of vision and visual consciousness. *Behavioral and Brain Sciences, 24*(5), 939–973.
Pettersson, R. (1982). Cultural differences in the perception of image and color in pictures. *ECTJ, 30*(1), 43–53.
Piccinini, G., & Shagrir, O. (2014). Foundations of computational neuroscience. *Current Opinion in Neurobiology, 25*, 25–30.
Pinker, S. A. (1998). *How the mind works*. Penguin Group.
Port, R. F., & van Gelder, T. (1995). *Mind as motion: Explorations in the dynamics of cognition*. MIT Press.
Putnam, H. (1981). *Reason, truth and history*. Cambridge University Press.
Pylyshyn, Z. W. (1984). *Computation and cognition*. MIT Press.
Ramstead, M. J. D., Kirchhoff, M. D., & Friston, K. J. (2019). A tale of two densities: Active inference is enactive inference. *Adaptive Behavior, 28*(4), 225–239.
Reber, A. S. (2019). *The first minds: Caterpillars, 'karyotes, and consciousness*. Oxford University Press.
Reed, E. S. (1996). *Encountering the world: Toward an ecological psychology*. Oxford University Press.
Rietveld, E., Denys, D., & Van Westen, M. (2018). Ecological-enactive cognition as engaging with a field of relevant affordances. In *The Oxford handbook of 4E cognition* (p. 41). Oxford: Oxford University Press.
Rizzolatti, G., & Craighero, L. (2004). The mirror-neuron system. *Annual Review of Neuroscience, 27*(1), 169–192. https://doi.org/10.1146/annurev.neuro.27.070203.144230
Rowlands, M. (2010). *The new science of the mind: From extended mind to embodied phenomenology*. MIT Press.
Rumelhart, D. E. (1997). The architecture of mind: A connectionist approach. In J. Haugeland (Ed.), *Mind design II: Philosophy, psychology, artificial intelligence* (pp. 205–232). MIT Press.
Rupert, R. (2009). *Cognitive systems and the extended mind*. Oxford University Press.

Sarkar, S. (1992). Models of reduction and categories of reductionism. *Synthese, 91*(3), 167–194.
Searle, J. (1983). *Intentionality: An essay in the philosophy of mind*. Cambridge University Press.
Searle, J. (1997). Minds, brains, and programs. In J. Haugeland (Ed.), *Mind design II: Philosophy, psychology, artificial intelligence* (pp. 183–204). MIT Press.
Shapiro, L. A. (2010). *Embodied cognition*. Taylor & Francis.
Simon, H. A. (1962). The architecture of complexity. *Proceedings of the American Philosophical Society, 106*(6), 467–482.
Simon, H. A. (1972). Theories of bounded rationality. *Decision and Organization, 1*(1), 161–176.
Smolensky, P. (1997). Connectionist modeling: Neural computation/mental connections. In J. Haugeland (Ed.), *Mind design II: Philosophy, psychology, artificial intelligence* (pp. 233–250). MIT Press.
Thagard, P. (2005). *Mind: Introduction to cognitive science* (2nd ed.). MIT Press.
Thelen, E., & Smith, L. (1994). *A dynamic systems approach to development: Applications*. MIT Press.
Thompson, E. (2007). *Mind in life: Biology, phenomenology, and the sciences of mind*. Harvard University Press.
Thompson, E. (2018). Review of "evolving enactivism: Basic minds meet content."
Thompson, E., & Stapleton, M. (2009). Making sense of sense-making: Reflections on enactive and extended mind theories. *Topoi, 28*(1), 23–30.
Thompson, E., & Varela, F. J. (2001). Radical embodiment: Neural dynamics and consciousness. *Trends in Cognitive Sciences, 5*(10), 418–425.
Todd, P. M., & Gigerenzer, G. (2003). Bounding rationality to the world. *Journal of Economic Psychology, 24*(2), 143–165.
Turing, A. M. (1997). Computing machinery and intelligence. In J. Haugeland (Ed.), *Mind design II: Philosophy, psychology, artificial intelligence* (pp. 29–56). MIT Press.
Turvey, M. T. (1992). Affordances and prospective control: An outline of the ontology. *Ecological Psychology, 4*(3), 173–187. https://doi.org/10.1207/s15326969eco0403_3
Turvey, M. T., & Carello, C. (1995). Some dynamical themes in perception and action. In R. F. Port & T. van Gelder (Eds.), *Mind as motion: Explorations in the dynamics of cognition* (pp. 373–401). MIT Press.
van Gelder, T. (1997). Dynamics and cognition. In J. Haugeland (Ed.), *Mind design II: Philosophy, psychology, artificial intelligence* (pp. 421–450). MIT Press.
Varela, F. J., Thompson, E., & Rosch, E. (2016). *The embodied mind: Cognitive science and human experience* (2nd ed.). MIT Press.
Wheeler, M. (2005). *Reconstructing the cognitive world: The next step*. MIT Press.

Wheeler, M. (2010). Minds, things and materiality. In L. Malafouris & C. Renfrew (Eds.), *The cognitive life of things: Recasting the boundaries of the mind* (pp. 29–38). McDonald Institute for Archaeological Research.

Wiese, W., & Metzinger, T. (2017). Vanilla PP for philosophers: A primer on predictive processing. In T. Metzinger & W. Wiese (Eds.), *Philosophy and predictive processing*. MIND Group.

Wilson, R. A., & Clark, A. (2009). How to situate cognition: Letting nature take its course. In M. Aydede & P. Robbins (Eds.), *The Cambridge handbook of situated cognition* (pp. 55–77). Cambridge University Press.

CHAPTER 3

Enactive Cognition: From Sensorimotor Interactions to Autonomy and Normative Behavior

Enactive cognition is one of the two fundamental branches of world-involving cognition. It explicitly rejects the two main dogmas of brain-centered cognitive science, which are neurocentrism and computationalism, and promises to overcome the problems of the mind–world dichotomy in scientific explanations of cognition. Neurocentrism uncritically assume this dichotomy and is therefore a problematic stance in cognitive science from both empirical and philosophical points of view. In contrast to brain-centered cognitive science, enactive cognition holds that cognitive processes take place across the brain–body–environment system. We can still use computational models to explain the activity of extended dynamical processes of cognition, but that does not necessarily mean cognition consists of information processing, as computationalism assumes. For enactivists, cognition is indeed enaction, even if an exact definition of this affirmation is sometimes lacking.

This chapter argues there are two main types of enactivism, and they may be distinguished by their interpretations of the claim that cognition is enaction. I will call these types *weak enactivism* and *strong enactivism*. Weak enactivism is built on the belief that cognition is rooted in sensorimotor correlations, and these correlations are established as a result of the interactional processes that agents and their environments sustain. Strong enactivism, on the other hand, sets the basis of cognition in the

M. A. Sepúlveda-Pedro, *Enactive Cognition in Place*, New Directions in Philosophy and Cognitive Science,
https://doi.org/10.1007/978-3-031-20282-7_3

constitutive autonomy of agents, and in their capacity to create and adapt their own norms of interaction with the environment. The theory of biological autonomy, the acknowledgment of subjectivity as an intrinsic dimension of cognition, and the notion of sense-making—all these make the enactive approach more robust, thereby offering a more accurate description of the systemic constitution of real cognitive agents (i.e., living agents).

In Sect. 3.1, I examine the foundational claims of enactive cognition in its seminal work, *The Embodied Mind*. This analysis will help us appreciate the original proposals of enactive cognition, and how the other variants of enactivism diverge from the original project. From there, I distinguish between weak and strong enactivism (Sect. 3.2). To clearly illustrate the differences between these two forms of enactivism, I will describe the main project of radical enactivism (Sect. 3.3). This form of enactivism is in many ways a solid research project, yet it nonetheless fails to live up to the original principles of enactive cognition. Finally, I will analyze the main ideas of the enactive approach, and specifically examine their use of the life and mind continuity thesis (Sect. 3.4), which, I will argue, is the key to revolutionizing the field of cognitive science and setting the new basis of world-involving cognition.

3.1 The Philosophical Foundations of Enactive Cognition

In *The Embodied Mind*, the founders of enactive cognitive science or enactive cognition, Francisco Varela, Evan Thompson, and Eleanor Rosch, made two central statements. The first is overall philosophical and tells us that cognition is a phenomenon that lacks absolute foundations and must be studied from both a first- and a third-person perspective. The second is rather scientific and aims to define the phenomenon of cognition as enaction. The following two subsections focus on the first statement, starting with the criticisms of Varela and colleagues to the main philosophical traditions that have influenced modern sciences of mind (Sect. 3.1.1), and continuing with the exposition of their own philosophical view (Sect. 3.1.2). Later, I analyze the two main interpretations of the concept of enaction (Sect. 3.2).

3.1.1 A World Without Egos, Egos Without Worlds

The Embodied Mind begins with criticisms of the two main philosophical tendencies in modern sciences of mind. Varela et al. (2016) called these two assumptions *realism* and *idealism*. I prefer to use the term *naïve objectivism* instead of realism to highlight that the problem of modern sciences of mind, for enactive cognition, is not to aim for some sort of objectivity in science, that is, the existence of objects that transcends our subjective experience. Instead, the problem is their blind adoption of the naïve attitude of many modern scientists and philosophers, which implies neglecting the role of our bodily, cultural, and historical immersion in describing and explaining all phenomena we experience. I also use the term *disembodied subjectivism* instead of idealism to highlight that although enactive cognition makes subjectivity a constitutive part of cognition, it is not a traditional (i.e., disembodied) form of subjectivity.

Naïve objectivism relies on three essential claims. The first claim is *ontological*: the objects of the world exist as such, that is, with intrinsic properties that are independent of our knowledge and experience of them. From this perspective, nature is a subject-independent reality (e.g., Bunge, 2006). The second claim is *epistemological* and amounts to saying that we can eventually know this subject-independent reality thanks to the inductive methods of science (Hacking, 1983; Mill, 2012). Since subjective experience is a phenomenon that others cannot observe, we require the experiential aspects accompanying cognition be excluded from the scientific study of cognition. Consequently, there is a third *methodological* assumption: we can eventually explain cognition employing the third-person perspective methodologies of science without considering the first-person perspective (Churchland, 1986).

The disembodied view of subjectivity that we generally associate with Descartes, by contrast, did not use empirical observation as its primary method for knowledge but is somewhat grounded on subjective thought. More specifically, this sort of subjectivism relies on deductive reasoning. Using a speculative and deductive method, Descartes (2008), the so-called father of this philosophical stance, criticized the fallibility of empirical knowledge and concluded that the most secure foundation of knowledge is our own ego. Hence, even if everything we perceive is illusory and what we think is false, it is still true that our thinking ego is being cheated, and it is unquestionable that our ego, as a thinking thing, as a *cogito*, exists. This existence is nonetheless abstract because, for Descartes, it is only the ego *as it is thinking* that exists. According to the standard interpretations of Cartesian thought, the cogito is a *disembodied ego*. Only in his rational capacities can we find the foundations of our knowledge of the external world (Rowlands, 2010).

For Varela et al. (2016), both naïve objectivism and disembodied subjectivism are misleading because they share a common error: the quest for secure foundations of knowledge. Neither the empirical world nor the cogito provides the solid ground that objectivists and subjectivists are looking for in their studies of mental phenomena. Hence, Varela and colleagues unveil the weaknesses of the philosophical views of modern sciences of the mind and propose a third way.

Concerning naïve objectivism, enactivists disclose the weakness of modern sciences of the mind when they neglect the relevance of subjectivity and lack the rigorous analysis of the first-person perspective.

Our study of cognition originates in and is continuously constrained by subjective experience. For instance, the causal explanations of perception, imagination, memory, decision-making, etc., are based on descriptions originating either in the experience of the cognitive scientist or in the reports of the experience of other subjects (eventually interpreted by the cognitive scientist). Scientists use these original "data" to elaborate theoretical hypotheses about mental phenomena causal processes. Since modern scientists have found significant correlations between brain activity and subjective experiences, mind scientists can infer a causal connection between reported subjective experiences and brain processes.

Nonetheless, subjective experiences are phenomena we cannot directly observe and are difficult to standardize. The reports and intuitions of different subjects, scientists, and philosophers are often variable, divergent, and contradictory. The empirical observations made in the controlled environments of laboratories, by contrast, are more reliable. Therefore, many scientists conclude that we must dispense subjective reports and base our causal explanations of mental phenomena on empirical data alone (e.g., Churchland, 1986).

The problems of measuring and controlling data provided by subjective experience prompted physiological approaches, like behaviorism, to neglect the importance of what takes place in the non-observational realms of mental phenomena. Cognitivism, the first and most classical form of brain-centered cognitive science, proposed instead to computationally model some of the non-observational aspects of mental phenomena (the functional elements). Computational models of cognition aimed to pass over the marshes of subjectivity, conjecturing formal models that more reliably correlated subjective reports to observed brain processes. Nonetheless, these computational models tended to forget their subjective origins and the importance of having continuous feedback from subjective experience to make these models more accurate (Jackendoff, 1987).

A disembodied subjectivism is no less problematic than naïve objectivism, in any case. According to subjectivist philosophies, the ego is what makes intelligible the external world. The causal laws we observe in the empirical world, for instance, are not part of nature as such but simply represent how our minds make sense of worldly events. The order might be psychological, as Hume (1748) suggested, or transcendental, as Kant (1998) proposed, but this does not change much. The nature of the ego is also more slippery than it seems *prima facie*. For Descartes, the existence of the cogito is a self-evident truth, but we cannot find the ego as an existent thing (as an object) anywhere.

If the cogito is not the foundation of experience as a substantial thing, it is still possible to claim that the ego is something like a transcendental self, that is, a condition of possibility of experience. Kant (1998), for instance, described some sort of transcendental ego as the unity of apperception. In Kant's philosophical system, this ego-like transcendental structure is the necessary condition for the unity and coherence of spatiotemporal experiences. Such an ego is, nonetheless, as with Descartes, a mere principle; as such, it cannot be experienced either. Kant's transcendental ego, too, thus appears as a pure abstraction, that is, as an ego that, enclosed in itself, finds itself only indirectly in the act of self-reflection (Varela et al., 2016, pp. 70–72).

Varela and his colleagues not only argue that it is impossible to find a concrete self or an ego in the speculation of Cartesian-like philosophies, but they also think that it is impossible to find it empirically. There is no evidence of a "central command" or a constant dynamic unity in brain processes that correlates to the unity of our different experiential acts. Instead, the brain exhibits a diversity of dynamic processes that vary according to the multiple conditions of the environment and the many tasks we must accomplish (Varela et al., 2016, pp. 72–79). That is, the unity of the self varies according to the nature of the practical tasks in the current circumstances. Therefore, the self is not a solid unity, independent of the dynamic interaction of an agent and the environment; its unity is somewhat contingent and depends on the coupling of the body and the environment (Varela et al., 2016, pp. 123–130).

Considering all the evidence against the foundationalist ambitions of objectivism and subjectivism, Varela and colleagues suggest an alternative approach to cognition that rejects grounding this phenomenon either in the objective world or in the subject's mind. Instead, we should recognize mind and cognition as phenomena that are *entre-deux,* a middle way, between these two opposites. The middle way is not exactly a third dimension between mind and world; rather, it is the original embodied dynamics in which self and world emerge as two poles of the same relational structure. Borrowing Fuchs's (2018) terminology for referring to the body as a dual unity where the first- and third-person perspectives converge, I will call this middle way *embodied subjectivity.*

3.1.2 Embodied Subjectivity

In *The Embodied Mind,* Varela and colleagues found the Madhyamaka school of Buddhism the most trustworthy source for understanding the middle way. However, in subsequent works, the philosophical discipline of phenomenology became the theoretical pillar of enactive cognition (Thompson, 2007). Since phenomenology is also highly relevant for my arguments in the following chapters, I am focusing here on some key ideas of this philosophical movement that have contributed to constructing the middle way of the enactive approach.

The thesis of embodied subjectivity begins with recognizing the existential condition of human beings that Heidegger (1962) called *being-in-the-world.* This being awakes to conscious reflection in a world already structured by relations of meaning or, more properly speaking, by what Merleau-Ponty (2012) called *sense.* For Merleau-Ponty, sense involves the particular orientation in which things appear before us in experience, most notably in perception. Sense is not intrinsic to things, though. It is akin to the practical and affective significance of things situated in specific contexts of action that we can call *horizons* (see Sect. 4.3). The significance of things and their horizons originates in the dynamic interaction of a living body and its surroundings. These surroundings are natural and physical, but, for humans at least, they are also social and cultural. Therefore, our lived experience of the world is not a reproduction or a representation of how the world is, independently of us; instead, it is the *disclosure* of our own immersion in the world we live in. As the starting point for the scientific study of cognition, the adoption of this existential condition described by phenomenologists radically transforms the project of cognitive science.

Phenomenology analyses our existential condition using a rigorous method to describe and analyze our lived experiences. Nevertheless, phenomenological descriptions are not about the contents of individual subjective experiences, as it is common to think. Phenomenology is somewhat concerned with the structural aspects or *invariants* of these experiences (Gallagher, 1997). Phenomenological descriptions are, in this sense, essential or eidetic, not psychological or introspective. To accomplish this sort of analysis, Husserl applied a strategy called the phenomenological *epoché* (Husserl, 1982, §32). The epoché puts on hold any judgment about the positive existence of the objects we experience, judgments we usually make in our everyday lives. Husserl calls this everyday attitude the *natural attitude* of experience (Husserl, 1982, §30). By utilizing the epoché, a new dimension of experience opens itself: We shift our attention from the things given in lived experience to *how* they are shown in lived experience. When doing so, we leave the natural attitude behind and adopt what Husserl calls the *phenomenological attitude* (Husserl, 1982, §50).

The natural attitude is not exclusive to our everyday experiences; it is also present in philosophy and science. Empiricism is the philosophical tradition that has most clearly (albeit only implicitly) endorsed this attitude as it grounds all kinds of knowledge on objects given *in* experience. Since empiricists recognized, like Descartes, that our senses are fallible, they transferred the foundations of knowledge to our capacity to infer the causal basis of phenomena from empirical observations (Mill, 2012). From this standpoint, therefore, causal theories of observable phenomena, rather than direct perceptual experiences, give us access to the intrinsic reality of the world. Husserl (1970) called this specific form of the natural attitude in modern sciences *the naturalistic attitude.*

The phenomenological method consists in systematically bracketing both our natural and naturalistic attitudes. After the epoché, the second fundamental step is *the phenomenological reduction* that asks us to adopt a transcendental analysis of the structures of experience (an analysis of how things are given within experience). That is, the epoché and the reduction altogether shift our attention from *what* is shown within lived experience (the empirical world) to the structures that make experience possible (the transcendental dimension of experience) (Zahavi, 2003).

On the foundations of Husserl's phenomenology, both Heidegger and Merleau-Ponty realized that phenomenology could not be the enclosure of a rational subject within its abstract thoughts because the subject is inevitably and permanently open to the world (Merleau-Ponty, 2012). Therefore, phenomenology is fundamentally the analysis of this existential condition.

The continuous and unavoidable opening of the embodied subject to the world makes phenomenology's methodology fallible (Merleau-Ponty, 2012). However, subjects' natural and naturalistic attitudes are not mere errors, defects, or weaknesses. They are instead the existential condition of beings in the world. The natural attitude corresponds to how we live every day when interacting with things and people around us in the most ordinary ways. The fundamental weakness of the natural and naturalistic attitudes lies elsewhere: It neglects the *facticity* of our existential situation as embodied and enculturated beings (Merleau-Ponty, 1964). Unless we pay attention to this situation, we risk falling prey to dogmatisms that can impose a single theory, ideology, or cultural form as universal models that produce universal and unquestionable truths.

Phenomenology engages in an exercise of self-reflection to avoid these dogmatisms. The goal is to identify the invariant structures that make up the constitutive entanglement of embodied (practical, affective, erotic, and expressive) subjects and their world. In the case of human beings, the embodiment is not a bare physical relationship between an isolated individual and the world. It is always also shaped by culture. Hence, our existential condition is embodied and enculturated.

Therefore, phenomenology's transcendental stance does not consist of abstracting the rational subject from the world in which she lives but in performing hermeneutics of facticity (Heidegger, 1962). That is, we do a self-interpretation of our embodied and enculturated existence, which is also the enactive approach ambition in cognitive science.

The enactive approach criticized the naturalistic attitude at the heart of neurocentric cognitive science, which completely disregards that embodied subjectivity is a condition for doing science. Before being a scientist, a person is a human being embedded in the horizons of the world she lives in. She adopts theoretical prejudices and observation methodologies, uses her age's academic and technological tools, and unavoidably biases her investigations by personal and cultural prejudices, etc. Her scientific labor thus necessarily arises within a pre-given and meaningful world that Husserl (1970) called the *lifeworld*. The human condition of cognitive scientists is what the authors of *The Embodied Mind* call a *fundamental circularity* because the scientists and the philosophers remain trapped in a circle of self-interpretation within the world.

Therefore, the first aim of enactive cognition in *The Embodied Mind* was to make cognitive scientists recognize this existential condition. This approach claims that we need a methodological analysis of experience to guide our scientific inquiries about mind and cognition to acknowledge cognitive agents' embodied and enculturated existence.

3.2 The Divergent Paths of Enactive Cognition

Upon reviewing the philosophical claims set out in *The Embodied Mind*, it should be clear that, from an enactive perspective, a scientific study of cognition should involve explanations that avoid positing absolute foundations. Therefore, if we claim that cognition is enaction, the roots of cognition should be contingent and labile, not predetermined.

This philosophical view, however, is admittedly still too general to serve as a definition of enaction, and it is also subject to different interpretations. To help clear this up, let us first recall that the word enaction has two different nuances of meaning. Enaction can refer to the *acting out* of a play or a story. In this sense, enacting something is like doing a *mise-en-scène* of a pre-given script, as when actors played *Caligula* at the Montreal *Théâtre du Nouveau Monde*. This enacting of Caligula is different from simply reading Camus' work in a book. A second sense of the word enaction refers to putting a new law into effect. Note that this kind of enactment produces a new set of constraints and rules for behavior in society. Thus, when the Canadian parliament made the public sale of Cannabis legal, it enacted a law that changed the habits of people living in Canada. With these two senses of the word in mind, we can now ask if the claim that cognition is enaction means that cognition arises from the action of agents as a result of predetermined guidelines or if it involves the creative emergence of new guides of behavior. Curiously, these two alternatives resemble two possible interpretations of Varela et al.'s claims in *The Embodied Mind*.

3.2.1 Weak Enactivism

The most common and loosest interpretation of the claim that cognition is enaction centers on the thesis that all forms of cognition are rooted in cycles of action and perception (e.g., Gallagher, 2017; Hutto & Myin, 2013; Myin & Degenaar, 2014; Noë, 2004; Rietveld et al., 2018). This conception of cognition as enaction contrasts with the claim of brain-centered cognitive science, according to which cognition is a computational function that simply processes information received from the outside world.

In the following two classical passages from *The Embodied Mind*, we can find support for this standard interpretation and what enaction means in this context:

> In the enactive program, we explicitly call into question the assumption—prevalent throughout cognitive science—that cognition consists of the representation of a world that is independent of our perceptual and cognitive capacities by a cognitive system that exists independent of the world (Varela et al., 2016, p. lxvi).
>
> We can now give a preliminary formulation of what we mean by *enaction*. In a nutshell, the enactive approach consists of two points: (1) perception consists in perceptually guided action and (2) cognitive structures emerge from the recurrent sensorimotor patterns that enable action to be perceptually guided (Varela et al., 2016 p. 173).

If we look attentively, the first passage describes the negative stand of enactive cognition against one specific aspect at the core of brain-centered cognitive science: representationalism. The second passage provides a positive definition of enaction as sensorimotor interactionism. *In this first interpretation, enaction consists of two sub-theses: Perception is guided by action, and any form of cognition is based on the same sensorimotor patterns that constitute perception.*

This first interpretation of enaction is not strictly wrong, nor does it necessarily contradict the philosophical thesis of embodied subjectivity. Nonetheless, this interpretation falls short of the philosophical commitments of Varela and his colleagues. The fact that cognition is rooted in sensorimotor correlations does not exclude the possibility that a set of pre-given properties of the world and the agents' bodies predetermine the constitution of their cycles of action and perception. That is, the definition still does not preclude one from locating the absolute foundations of cognition in the constitution of the subject independent world.

Although the concrete action of agents is needed *to enact* the relational field of movement and sensation, the characteristics of the relational field are fundamentally pre-given, like the script of a play. This first interpretation of the claim that cognition is enaction in *The Embodied Mind* thus corresponds with the first meaning of the word enaction noted above.

I call the type of enactivism derived from this interpretation *weak* because they make a partial and superficial reading of the claims of Varela and his colleagues in *The Embodied Mind*. Moreover, if our view of cognition depends on no more than sensorimotor interactions, we risk missing the essential features of cognition and cognitive agents. Later, I show these weaknesses exemplified in radical enactivism (Sect. 3.3).

3.2.2 Strong Enactivism

For the founders of enactive cognition, perception and cognition do not simply emerge from a physical system that maintains efficient sensorimotor interactions with the environment. *For a system to be cognitive, its interactions with its surroundings must be relevant or significant for the system itself.* We cannot just preprogram significance, as held by older models of cognitivist artificial intelligence, nor can it be predesigned in the body type of agents, as with the models of embodied artificial intelligence (see, e.g., Sects. 2.2.1 and 2.3.3). The significance of sensorimotor interactions must be relevant for the dynamic constitution of the cognitive agent and the patterns of behavior this agent has enacted in its previous interactions with the environment. That is, *the sensorimotor interactions of cognitive agents are normative* (they are correct or incorrect), but the norm at work in these interactions cannot be given in advance or independently of the concrete existence of the agent.

On the contrary, *it is in the action of agents that interactional norms are enacted*. This conception of enaction gets closer to the second meaning of the word enaction that we read above, that is, the act of putting laws that guide and constrain the behavior of individuals in society into effect. Nevertheless, there is a difference, for, in this case, subjects enact themselves the laws (or norms) and not an entity that sits hierarchically "over" them. This definition helps us see that the hallmark of a strong enactivism equates enaction to the act of *bringing forth a world of significance based on the autonomy* that characterizes the constitution of a cognitive agent (Varela et al., 2016, p. 155). Let me explain this idea more clearly.

Enacting or bringing forth a domain of significance means that, from an environment that is chaotic or structureless, a cognitive system can produce regular patterns of interaction and become responsive to a specific set of features in the environment. The example of the Boolean cellular automata named Bittorio shows this phenomenon (Varela et al., 2016, p. 150–157).

Bittorio is a ring with eight modules or cells with a value of 1 or 0. These rings interact with a chaotic environment of random values (also of 1 or 0), thereby altering the emergent patterns of Bittorio's cells. After enough time interacting with the environment, Bittorio self-organizes and becomes responsive only to some patterns of interaction, whereas other patterns do not alter its structure.

One of the most relevant aspects of this example is that Bittorio is not preprogrammed to accomplish its task. Instead, what makes Bittorio capable of generating this "structural coupling" with the environment is its "autonomous organization." The authors of this book use autonomy as a synonym for *operational closure*, which means that the activity of the different components of a system is interconnected. In this case, a change to the state of one of the cells of Bittotrio alters the whole organization of the system. However, the entire organization of the system will depend on how this alteration is related to the current states of all other cells. The organization the automata acquires over time (autonomous organization), and the responsivity of this system to more specific inputs (structural coupling) is a process of self-organization, and it remains open to structural changes. In Varela's interpretation, the relevant patterns of interaction *become* Bittorio's world. It does not mean Bittorio has an experience of its environment. It only means that a system with autonomy can create order from a chaotic background, but this order co-emerges with the systemic organization of the system.

Therefore, the theory of autonomy is central to defining cognition as enaction in a more profound sense than weak enactivism. The theory of autonomy makes enaction fundamentally an act of bringing forth a domain of significance that is not previously given, either in the cognitive system or in the intrinsic features of the world but is rather enacted in the history of interaction between an agent and its environment.

We can now appreciate the intimate connection between the philosophical claims of embodied subjectivity and the scientific claims of the theory of autonomy from the perspective of strong enactivism. We saw earlier that embodied subjectivity implies the philosophical thesis that we cannot understand cognitive phenomena in purely objective or subjective terms. In contrast, the middle way more appropriately describes the dynamic entanglement of the body and the world. The theory of autonomy embodies the middle way, grounding cognition in the interaction of the agent–environment system and not in the pre-given constitution of the two components of this system. Therefore, claiming that cognition is

Table 3.1 Summary of the main differences between weak and strong enactivism

	Weak enactivism	*Strong enactivism*
Meaning of enaction in English	Acting out.	Law (or norm) creation.
Meaning of cognition as enaction	Cognition is grounded on sensorimotor correlations.	Cognition brings forth a world of significance for autonomous agents.
Philosophical foundations	The relational sensorimotor field of the agent–environment system builds the foundations of cognition.	There are no absolute foundations of cognition.

enaction implies much more than changing a representational model for a sensorimotor one (see Table 3.1). We also need to create scientific models that acknowledge the contingent existence of bodily agents laying down their paths while walking (Varela et al., 2016, p. 239).

The following section exposes how radical enactivism embodies weak enactivism. From there, I will consider the problems of radical enactivism and argue that only by adhering to a strong enactivism, like the one of the enactive approach and its mind and life continuity thesis (Sect. 3.4), we can avoid these problems.

3.3 Radical Enactivism as Weak Enactivism

There are currently three fundamental branches of enactive cognitive science: the enactive approach (Di Paolo et al., 2017, 2018; Thompson, 2007), more commonly labeled autopoietic enactivism; sensorimotor enactivism (Degenaar & O'Regan, 2017; Noë, 2004; O'Regan & Noë, 2001); and radical enactivism (Hutto & Myin, 2013, 2017). As mentioned, the enactive approach comprises the original project of enactive cognition, and, as I will show in Sect. 3.4, it continues the construction of the view that I have recognized as strong enactivism. Ten years after the publication of *The Embodied Mind*, the second branch of enactive cognition appeared as *the theory of sensorimotor contingencies* (O'Regan & Noë, 2001) but is nowadays known as sensorimotor enactivism. The theory of sensorimotor contingencies is probably the most influential and grown approach to perception in enactive cognition. It focuses on explaining the dependency of the contents of perceptual experience on laws that govern sensorimotor correlations and offers strong support to the claim that action guides perception (Noë, 2004). Indeed, some of the proposals of sensorimotor enactivism have been useful in advancing the enactive approach to sensorimotor habits (Di Paolo et al., 2017). The original theory of sensorimotor contingencies, however, makes no explicit reference to the theory of autonomy, and autonomy does not seem relevant for the study of perception according to the contemporary supporters of sensorimotor enactivism (Degenaar & O'Regan, 2017). Hence, sensorimotor enactivism is the first case of weak enactivism in the history of enactive cognition. The third main branch of enactive cognitive science is radical enactivism, which is nowadays the most robust and complete project of enactive cognition after the enactive approach. For this reason, I analyze its main proposals to exemplify weak enactivism and show why the strong enactivism of the enactive approach offers a more promising view of world-involving cognition.

In this section, I first survey the main arguments of radical enactivism against representationalism in cognitive science (Sect. 3.3.1). Later, I show the theoretical bridges radical enactivists have built between their nonrepresentational approach to cognition and evolutionary biology (Sect. 3.3.2). Despite the undeniable contributions of radical enactivism to enactive cognition, it represents a case of weak enactivism that causes two important issues.

First, radical enactivism, like other forms of embodied cognition, has no criterion to distinguish between truly intelligent beings and efficient sensorimotor machines (Sect. 3.3.3). Even if we appeal to dynamical systems theory, which is one of the theoretical pillars of the whole field of world-involving cognition, we cannot distinguish between bare self-organizing systems and self-organizing cognitive systems (Sect. 3.4.4). In contrast, the autopoietic theory, from which the theory of biological autonomy of the enactive approach was born, already acknowledges this systemic difference (Sect. 3.3.4).

Second, radical enactivism endorses a conception of living organisms akin to the one of brain-centered cognitive science, an antagonist to the theory of biological autonomy. Rejecting the theory of biological autonomy, nonetheless, lets unexplained one essential feature of living organisms: agency. Consequently, radical enactivism fails to explain basic forms of cognition. I address this issue later in Sect. 3.4.

3.3.1 Anti-Representationalism and Teleofunctionalism

Radical enactivism was born as an approach that primarily focused on rejecting representationalism. For radical enactivism, representations are vehicles of contentful information. That is, they are internal brain states with conditions of correctness. The conditions of correctness of a representational state, such as a perceptual state, is a truth-condition. However, for radical enactivism, this is not an accurate description of what happens in basic forms of action and perception; whence the need to develop a sustained critique.

The main argument of radical enactivists against representationalism is that the information used by basic cognitive systems does not need to be contentful (Hutto & Myin, 2013). Instead, information at this level is the lawful covariation between different world facts. Covariations do not need to be represented because they already exist in nature. The number of rings of a tree and the number of times the Earth travels around the sun covary independently of our knowledge of them. The two phenomena, nonetheless, are not informational until there is an agent capable of disclosing the covariation between the two events (Hutto & Myin, 2013, pp. 63–71). For radical enactivism, this is what so-called basic minds do.

Fundamentally, basic minds are sensorimotor systems that can be us when we perform elementary sensorimotor tasks, such as walking, swimming, or chopping wood. Basic minds can also be animals or artificial intelligence robots like Brook's (Hutto & Myin, 2013, pp. 39–46). Radical enactivism affirms that there are lawful correlations between movement and sensory feedback. Given these correlations, the environment provides information thanks to the lawful covariation of movement and sensory changes. Basic minds do not need to encode anything from the outside to enact this information. They only need to be sensitive to the relevant informational covariations (Hutto & Myin, 2013, p.68). Therefore, the conclusion is that *basic cognitive systems exploit the information already available in the environment*. In this light, internal representations appear superfluous.

Despite the convergences of radical enactivism with various claims we find in the work of Varela and colleagues, radical enactivism never endorsed the theory of autonomy or the thesis of embodied subjectivity. Instead, radical enactivism bases its natural explanations of basic cognition on views of biology that contradict the claim that cognitive agents are autonomous systems, as strong enactivism implies.

Radical enactivism appeals to biological teleofunctionalism to explain the natural origins of non-contentful sensorimotor information (Hutto & Myin, 2013, pp. 71–82). For radical enactivism, the ability of living organisms to exploit environmental information is not directly caused by purely physical determinations. Instead, for radical enactivism, sensorimotor interactions of living organisms are normative. These norms are functional norms that involve the suitable accomplishment of a predeterminate biological function (Hutto & Myin, 2017, p. 116).

Biological functions refer to the usefulness that traits and behaviors of living organisms have or have had in the past for the survival of their species. These functions possess the *telos* of stably maintaining a species' fitness (successful reproduction rates). It is hence a *teleofunctionalism*. Organisms do not necessarily use these functions purposively. However, we can observe externally that a trait or behavior of an organism has a determinate use in terms of its survival and reproduction.

For philosophers of mind like Ruth Millikan (1984) and Fred Dretske (1986), teleofunctionalism can explain the natural origin of internal representations. This approach is known as teleosemantics. This perspective explains how mental contents correlate normatively to the world's facts. For teleosemantics, organisms are efficient in representing the outside world because their representational functions have been helpful for the preservation of their species (Godfrey-Smith, 2006). Evolutionary processes of natural selection are, therefore, the causal origin of representationalist functions.

3.3.2 *The Blind Watchmaker*

Radical enactivism adopts a stance akin to teleosemantics for explaining the origins of basic cognition, although it puts aside representationalism. Following Fodor's (1990) criticisms of teleosemantics, radical enactivism holds that what matters for preserving life and reproduction is not knowledge of facts of the world as such (true statements). Instead, for radical enactivism, it is the efficient exploitation of the information already available in the environment that counts as phylogenetically valuable for species survival (Hutto & Myin, 2013, p. 81).

For representational theories of mind like teleosemantics, intentionality produces internal representations of the external world. These theories endorse a Brentanian type of intentionality (see Sect. 2.3). On the other hand, radical enactivism proposes to think of exploiting environmental information (as covariation) in terms of a non-Brentanian form of intentionality that its supporters call *Ur-intentionality* (Hutto & Satne, 2015). This intentionality, I insist, implies a normative relation between cognitive systems and their environments. The adjective *normative* indicates that there are *right* (and *wrong*) accomplishments of biological functions determined by natural selection processes (Hutto & Myin, 2017). Radical enactivism describes, therefore, *functional normativity*.

Once it has explained the natural origins of non-contentful forms of intentionality, the second challenge for radical enactivism was understanding the natural origins of content (Hutto & Satne, 2015). Some forms of cognition, like thinking and reasoning, depend on propositional attitudes that refer to states of affairs in the world. These forms of cognition need conditions of correction that Ur-intentionality cannot provide. Here is when Wittgenstein comes in.

In accord with Wittgenstein, radical enactivism holds that sociocultural practices, especially linguistic practices, in contrast to natural processes, involve truth conditions, but these conditions depend on public norms (Hutto & Myin, 2017). That is, it is only in the public domain that statements about the facts of the world can be evaluated as true or false. In this regard, it is essential to note that the emergence of complex forms of human cognition also depends on evolutionary processes. Evolution and natural selection cause the emergence of sociocultural practices and linguistic exchanges in humans. Therefore, the origin of cognition and its development from basic levels to higher ones is explained phylogenetically, and evolutionary processes become the fundamental pillar of natural explanations of radical enactivism.

Biological functionalism is an explanation of life that strongly resonates with cognitivism's causal explanations because both equate living agents to artificial machines. Mechanistic and functionalist conceptions of life have their own historical tradition in theoretical biology. In light of the emergence of Newton's physics and Descartes' mechanistic philosophy of nature, at the beginning of the nineteenth century, William Paley (2009) thought that nature was like a giant clockwork machine that functioned according to a purpose set by God, who was akin to a watchmaker. In this picture, the order of the natural world is given by a previous design. Therefore, a metaphysical entity predetermines an organism's behavior.

Charles Darwin's (Darwin, 1859) theory of evolution made this conception of nature hard to support because the idea of the evolution of species fundamentally depends on natural selection. This process requires that random changes occur in nature (e.g., gene mutation), thus allowing species to adapt to the changing circumstances of the environment. Working under the paradigm of *The Modern Synthesis*, that is, the synthesis of Darwin's theory of evolution and Mendel's laws of genetics, contemporary biologists like Richard Dawkins (1986) took up the watchmaker metaphor and claimed that nature played the role of a blind watchmaker. Namely, even though no pre-given teleology exists in nature, there is yet a machine-like functioning at work. This machine is no longer the clockwork machine envisioned by scientists and philosophers of the seventeenth century; instead, it incorporates randomness, chaos, and overall temporal evolution. There are nonetheless lawful processes that constrain chaos and randomness. In the case of life, this process is natural selection.

Natural selection implies a process through which biological lineages of phenotypes (species) survive, maintaining a steady reproduction rate (i.e., fitness) over broad periods of time. For the supporters of The Modern Synthesis, adaptation is the fundamental mechanism of natural selection (Lewens, 2007). This process involves the random variation of genotypes (genetic information) that produce variants in the phenotypes of species. Some of these variants are more successful than others in their fitness because they are better adapted to the conditions of the environment. In the long run, phenotypes with better fitness become predominant in their species. This view usually conceives organisms as an ensemble of functional systems that have been selected phylogenetically (Gould & Lewontin, 1979). Every functional system, including an organism's behavior, contributes to its species' survival and fitness (Cosmides & Tooby, 1987; Dawkins, 1982). It is precisely the idea behind teleofunctionalism.

From the standpoint of radical enactivism, the capacity to exploit information as covariation is one of the adaptation processes of organisms. If this is the case, the blind watchmaker or, as Millikan (1984) prefers to name it, Mother Nature, is the only one responsible for the design of living and cognitive systems.

3.3.3 *The Missing Mark of the Cognitive in Radical Enactivism*

Radical enactivism has construed a solid approach to cognition. Its non-representational theory of information and its support from teleofunctionalism have gained credibility in the community of cognitive scientists for a non-neurocentric study of cognition. Nevertheless, all these benefits depend on weak enactivist claims, specifically on locating the roots of cognition in sensorimotor interactions, but without reference to the early theory of autonomy of Varela et al. (2016) and without assuming that we need non-absolute foundations to explain cognition. For radical enactivism, subject-independent processes, biological and physical, are the causal bases of basic cognition. Why do we need to look for more than weak enactivism and a more robust version of enactive cognition?

There are three important reasons to look for something more than what radical enactivists have to offer. First, radical enactivism cannot distinguish between pure mechanical systems exhibiting intelligent-like behavior and natural intelligent agents. Second, sensorimotor interactions may very well be helpful to enact information as covariation, while coordinating the body and the environment thanks to self-organization processes. However, from a systemic stance, there is no clear criterion to distinguish between two self-organizing systems that coordinate (a common phenomenon of nature) and systems that use this coordination to transform their own dynamics and the coupling they maintain with other systems. Third, there is no account of the agency of cognitive systems, at least not at the basic level of cognition. Agency seems to be a characteristic of living systems, systems we know are undoubtedly cognitive, but basic minds do not imply this feature. This section focuses on the analysis of the first issue of radical enactivism, Sect. 3.3.4 focuses on the second. Later, after reviewing the enactive approach mind and life continuity thesis, in Sect. 3.4.4, I will address the problems of biological agency.

Despite the significant support that radical enactivists find in theories of biological teleofunctionalism for their theory of Ur-intentionality, their descriptions of basic minds and basic cognition do not exclude the possibility that artificial systems can exploit information as covariation in the same way biological systems do. From the standpoint of enactive cognition, this is, in principle, not an issue for radical enactivism since Varela et al. (2016) also recognized the similarity between Brooks' embodied artificial intelligence and the scientific claims of enactive cognition. For Varela and colleagues, however, the early theory of autonomy can already posit a difference between the artificial systems created by embodied artificial intelligence and those of living systems. This difference is more evident in the theory of biological autonomy advanced by the enactive approach. The differentiation between artificial and biological cognitive agents is vital for cognitive science because it helps us distinguish between those systems that only *seem* intelligent and those that are genuinely cognitive agents. Kenneth Aizawa (2014) has made critical remarks about this cluster of issues.

Aizawa thinks that brain-centered cognitive science and enactive cognition describe different phenomena when they talk about cognition. Whereas brain-centered cognitive science traditionally focuses on complex forms of cognition like reasoning and thinking, enactive cognition has been more attentive to the basic forms of action and perception. Cognition, for brain-centered cognitive science, thus consists in representing the outside world to solve problems and, for enactive cognitive science, essentially consists of intelligent and/or viable behavior.

The adjectives 'intelligent' and 'viable' are nonetheless tricky. For brain-centered cognitive science, intelligent behavior implies cognition because the efficient accomplishment of motor actions involves computational information processing. Brooks' models, however, are more successful than neurocentric models, and they do not need to accomplish any information processing. This embodied artificial intelligence model maintains a continuous and "efficient" or viable sensorimotor coordination with the environment. Where then does the intelligence of these embodied mobile robots lie?

The intelligence of robots might be seen in the efficiency of their behavior, that is, in sustaining viable sensorimotor interactions. However, robots are, in the end, sophisticated machines designed to accomplish sensorimotor tasks efficiently. The fridge in my kitchen is undoubtedly a less sophisticated machine than a mobile robot, but it is also a highly efficient machine for maintaining food at a specific temperature. Its efficiency does not make my fridge intelligent. Without the specification that cognition consists of information processing, it becomes harder to define what is cognitive and what is not. The upshot is clear: *Claiming that (basic) cognition consists of efficient sensorimotor interactions is not enough.*

Radical enactivism is not only incapable of distinguishing theoretically between truly cognitive systems (as far as we know, many living systems) and cognitive-like systems (e.g., humanoids). It is also incapable of distinguishing *systemically* between merely self-organizing systems (e.g., whirlpools) and cognitive systems as such. To appreciate this point better, we need first to consider dynamical systems theory's role in embodied and enactive cognition.

3.3.4 The Missing Mark of the Living in Radical Enactivism

Dynamical systems theory has been one of the pillars of enactive cognition since this theory affords mathematical models to cognitive scientists for understanding the dynamical, normative behavior of the brain–body–environment system without implying that internal representations are involved in these processes (van Gelder, 1997).

A dynamical system is one that continuously evolves according to different variables. We can model almost every physical system as a dynamical system; a tornado, a steam engine, a living organism, a flock of birds, and the human brain are just some examples. In cognitive science, multiple brain, bodily, and environmental processes, causally relevant for accomplishing cognitive activities, can be modeled as dynamical systems. Moreover, it is possible to model the coupling of all these processes and claim that cognitive systems cut across the brain, the body, and the environment (Clark, 1997).

A dynamical system evolves according to multiple variables that affect the system's behavior to some degree. The temporal evolution of a system usually generates recurrent patterns of behavior that work as a set of internal constraints. These constraints partially determine the system's behavior and how the system responds to new variables. Since these constraints evolve continuously due to changes in the value of the variables, the system's behavior remains open to new changes and the emergence of new constraints (Thelen, 1995). A dynamical system is thus a temporally open-ended system. However, the open-ended nature of dynamical systems is not only diachronic (open to changes of behavior in the future) but also synchronic (a behavior coupled to the behavior of other dynamical systems). If two or more dynamical systems mutually constrain their behavior, they become coupled and share a common history of temporal evolution (van Gelder, 1997).

One remarkable aspect of dynamical systems theory for cognitive science is that we do not need to define external variables of the system as inputs that need to be encoded in discrete units that the system will process as vehicles of information. Somewhat, these inputs will alter the system's dynamic behavior according to the already established constraints of the system. The system's responses (outputs) do not need specific action commands but can be typical system reactions according to its dynamical configuration. Hence, dynamical systems theory makes room for nonrepresentational models of cognition.

For radical embodied cognition, including radical enactivism, dynamical systems theory is a tool that allows us, as external observers, to understand how dynamic processes of the brain, the body, and the environment system are coupled (Hutto & Myin, 2013). This is the most recurrent use of dynamical systems theory in world-involving cognition.

From a dynamical perspective, it is impossible to distinguish between two basic physical systems that exhibit a coordinated behavior while interacting dynamically and systems that use some form of cognition to interact appropriately with the environment. A Beer's robot, for instance, can recognize different shapes in figures and move efficiently to avoid diamonds and catch circles. The robot learned this behavior thanks to evolutionary algorithms implemented in a connectionist network (robot's brain). A dynamical system model can capture the self-organization processes that lead the robot to perform efficient sensorimotor interactions (Beer, 2000). Externally, we can observe analogies between this artificial system and a living sensorimotor system, but it does not mean that Beer's robot is genuinely a cognitive agent.

From a strictly formal standpoint, and putting aside the system's complexity, there is no strict differentiation between the type of coordination between an artificial agent and its surroundings, and the emergent coordination, after a while, of two pendulums that swing together on a movable surface. A dynamical systems description to distinguish between a cognitive and a noncognitive system is therefore necessary, but the radical enactivism description of basic minds does not imply it. The strong enactivism of the enactive approach, by contrast, entails this distinction.

Strong enactivism, as we saw above, involves a theory of autonomy. After *The Embodied Mind*, this theory slowly evolved into the theory of biological autonomy (Thompson, 2007), which posits that the primary organization of life (an autonomous organization) is a requirement for cognition. The first step toward a full-fledged theory of biological autonomy was elaborating the concept of autopoiesis, advanced by Varela and the Chilean biologist Humberto Maturana (1980). They claimed that what fundamentally characterize living organisms are their *self-producing* and *self-maintenance* organizational features. Hence, living systems are a network of causally interdependent processes that allow a system to maintain a specific organization for a certain period, which means that living organisms possess an organization that keeps them alive. Part of the processes that physically constitute the body of an organism is usually the construction of *a semipermeable boundary*. This boundary separates autopoietic processes from the rest of the environment but allows the exchange of matter and energy that an organism requires for its autopoietic activities.

The paradigmatic case of autopoiesis is the living cell (Thompson, 2007, p. 98). A cell is a living organism that constitutes its own operational realm by building a semipermeable boundary, or membrane, that separates its autopoietic processes from the rest of the environment. Inside this membrane, an interconnected network of processes occurs. These processes maintain the organization of the cell and further produce its physical constituents from resources provided by the environment. The membrane, then, is a semipermeable boundary, allowing for the controlled exchange of matter and energy between the cell and the environment. Indeed, only through this exchange can the cell maintain its autopoietic organization (Weber & Varela, 2002).

The metabolic needs of an organism, therefore, guide or provide the norm for the organism's sense of its behavior. Thus, bacteria (*E. coli*) tend to move toward an environment with a high glucose gradient because they find this place more appropriate for their metabolic needs. For this reason, a living organism, as an autopoietic system, constitutes what the ethologist Jacob von Uexküll described as an *Umwelt* (Thompson, 2007).

An *Umwelt* is the meaningful world a living organism enacts, similar to Bittorio's enactment of recurrent patterns of interaction. However, unlike Bittorio, living organisms distinguish the features of the environment according to their own existential concerns (to stay alive). In the case of living beings, therefore, enacting a domain of significance involves bringing forth differences that make a difference (Thompson, 2007, p. 57).

The theory of autopoiesis is thus fundamentally different from radical enactivism. For the holders of the theory of autopoiesis, living organisms are not blind mechanisms that behave according to functions completely predetermined by their phylogenetic past. Instead, they are agents that need to act according to a basic norm that is intrinsic to their existence: They must stay alive.

Any form of enactive cognition that adopts the route of strong enactivism must therefore endorse the claim that autonomy is an intrinsic feature of cognitive systems. Cognitive systems enact worlds of meaning or, in other words, enact their normative domain of interactions with the environment. Enacting their own norms based on the systemic composition of their bodies is fundamental to explaining agency, which is an essential characteristic of living and cognitive beings. Agency is precisely what the theory of biological autonomy of the enactive approach explains and what sets the basis for understanding the emergence of cognition from the groundless grounds of life.

3.4 The Enactive Approach as Strong Enactivism

The early theory of autonomy of the enactive approach was not only controversial and provocative for the orthodoxy of cognitive science but also problematic for those sympathetic to this theory. Describing computer programs capable of enacting worlds was a problematic way of thinking about real agents capable of enacting living domains of significance (Bourgine & Stewart, 2004). The theory of autopoiesis as the basic description of autonomy in real living systems was also unsatisfactory for describing the concrete temporal unfolding of life (Bitbol & Luisi, 2004; Di Paolo, 2005). Considering all these problems and controversies, the supporters of the enactive approach elaborated two crucial concepts: the theory of biological autonomy and the definition of cognition as a form of sense-making. This section reviews the main features of these two concepts.

3.4.1 Biological Autonomy

The definition of life and cognition as autonomous processes, described by the theory of autopoiesis and the early theory of autonomy, respectively, are finally synthesized and refined by Thompson (2007) into what we now know as *the life and mind deep continuity thesis* (Froese & Di Paolo, 2009). According to this thesis, life becomes a necessary condition for cognition, and cognition becomes a special form of bodily interaction that resembles the fundamental interactions at play in the most rudimentary forms of life.

The starting point for describing the continuity of life and mind is the systemic description of living systems as *autonomous systems in precarious conditions with adaptive behavior* (henceforth *autonomous systems*) (Thompson, 2007). This basic description is later reproduced at multiple levels of interactions (sensorimotor, intercorporeal, and linguistic) in an agent–environment system.

Defining living systems as autonomous follows from the theory of autopoiesis. The notion of autonomous systems is nonetheless a more general definition than autopoietic systems. While it is true that all autopoietic systems are autonomous, not all autonomous systems are autopoietic. Autopoietic systems self-maintain and self-produce their systemic organization and the physical boundaries that separate this organization from the rest of the environment. Nonetheless, autonomous systems self-distinguish themselves from the rest of the environment without constructing actual physical boundaries. The nervous and the immune system are two examples of autonomous but non-autopoietic systems (Varela, 2000, pp. 51–53). Therefore, the distinguishing feature of autonomous systems is not self-production but *operational closure.*

Operational closure describes the interdependency of a network of processes necessary to sustain the dynamic organization of a system over time (Di Paolo & Thompson, 2014). The dynamic organization of the system also depends on causal processes that happen beyond the organizational boundaries of the network. However, such processes do not rely on the activity of the network. Hence, there is an organizational but not necessarily a spatial distinction between what is "in" and outside the autonomous system. For this reason, we can rightfully describe as autonomous many other living systems apart from unicellular organisms (the paradigmatic example of autopoiesis).

The shift from the theory of autopoiesis to the theory of biological autonomy also involves a more sophisticated description of the interactions a living agent maintains with the environment. In this regard, there is a marked difference between the early days of the theory of autopoiesis and the period that Barrett (2017) calls the normative turn of the enactive approach. This turn starts with Weber and Varela's (2002) appeal to the phenomenological work of Hans Jonas, who claimed that living beings exhibit an intrinsic teleology.

For Jonas, this intrinsic teleology is not a mere attribution of a property of living beings from the point of view of an observer but a feature inherent to the existential condition of living beings who must constantly escape their precarious condition. The laws of thermodynamics state that all physical systems tend to lose their systemic organization, which means that over time, they face disintegration. In the case of living beings, this means the constant threat of death. Organisms can delay tendency through interactions with the environment, and metabolism is the most fundamental of these interactions. In light of this, Jonas (1966) claimed that living beings do not behave according to "external" causal determinations but according to their most fundamental norm: that of their self-preservation. For this reason, the mother of all values is the affirmation of life (Weber & Varela, 2002).

The normative turn, however, also implies a shift in how to conceive this fundamental norm. Rather than being a positive affirmation of life, self-preservation of life is ***a double negation*** of its existential condition (Di Paolo & Thompson, 2014). The bodily existence of living organisms has a natural tendency to death (a negation of life), but to be alive implies constantly moving away from this natural tendency (a negation of a negation). Life is, therefore, a frustrated suicide (Di Paolo, 2009, p. 16) and reveals a permanent tension within organisms between these two opposite tendencies. It is not banal this clarification because it shows the sort of normativity at work in the framework of the enactive approach. One of the consequences of this claim is that norms of life are norms of viability instead of optimality (Thompson, 2007, p. 206). As long as the tension within the organism maintains viable the continuity of the system, the structural coupling with the environment remains stable, and the norm is satisfied. There is no optimal adaptation of the organism to the environment because there are no optimal (a priori) conditions the organism must satisfy.

It is important to note that Jonas' description of precariousness in living beings does not suffice to define the existential condition of life (Barandiaran & Moreno, 2008). According to the laws of thermodynamics, it is not only living systems that find themselves in precarious conditions and threat by thermodynamical equilibrium but all physical systems. Nonliving self-organizing systems like whirlpools and tornados also avoid entropy temporarily. Hence, we can say that precariousness and self-preservation by themselves do not define life.

In response to this issue, the enactive approach recognizes that only living beings can establish an ***interactional asymmetry*** with their surroundings, a feature that mere physical self-organizing systems lack (Barandiaran et al., 2009). Thus, living organisms can modulate and regulate their coupling with their soundings to produce effective ways to escape systemic disintegration, thereby genuinely becoming agents.

In this respect, when Di Paolo (2005) adds the feature of ***adaptivity*** to Varela's theory of biological autonomy, the normative turn is finally accomplished (Barrett, 2017) because the agent's capacity to modulate its coupling with the environment significantly determines the norms of the interactional coupling. Consequently, the environment is enacted as a field of opportunities for altering the systemic conditions of living systems.

3.4.2 Sense-Making

A living agent's lived world is a domain relevant to its own existential concerns. What we can see as a region of a meaningless domain of physics appears for a living agent as a meaningful world or an *Umwelt* (Thompson, 2007). The enactment of this domain of significance is what the enactive approach calls *sense-making*. Enactivists paradigmatically illustrate this phenomenon with experiments where *E. coli* bacteria are behaviorally responsive to high concentrations of glucose because it is valuable for the metabolic needs of these bacteria. They are, by contrast, irresponsive to other chemical components that do not affect their autonomous organization. Properly speaking, sense-making is not a cognitive or at least not a purely cognitive act that informs living agents about the external conditions of the environment. It is instead an affective sentience of the conditions of equilibrium of the agent's body in relation to the conditions of the environment. Let me explain this idea more clearly.

When the *E. coli* environment is rich in glucose, the presence of this chemical compound in a specific spatial area alters the bodily equilibrium of bacteria. Alterations in the dynamical bodily processes of *E. coli* change the type of movement their flagella perform, so they can hang around in that area instead of passing by and continue their movement to other regions (Egbert et al., 2010). Consequently, there is a bodily responsiveness to the conditions of the environment.

For the enactive approach, the same type of bodily affectivity happens for all living beings, from bacteria to humans (Colombetti & Thompson, 2008). We must be aware that basic biological affectivity is not necessarily like the complex emotions we live as human beings (Colombetti, 2014). The basic affectivity of life is the bodily tension that organisms undergo due to states of organizational disequilibrium. The relevant aspects of the environment for autopoietic processes are then lived as affective bodily tensions (Colombetti, 2018).

This basic affectivity of life is akin to the affective state of humans, described by Fuchs (2012) as "the feeling of being alive," which implies all the brain activity related to the basic regulatory processes of the body (Colombetti & Thompson, 2005). This feeling of being alive is essential for sense-making and cognition (Fuchs, 2018). The conclusion to draw from all this is obvious: To understand sense-making, we must stop thinking about the enactment of *Umwelten* in purely cognitive terms. We must consider this enactment correlated to all living beings' basic bodily affectivity. Therefore, this basic form of sense-making involves what we might call *bio-affectivity*.

The idea of bacteria having affections like ours is undoubtedly controversial. Nonetheless, the enactive approach has been clear about the incommensurability of experiences due to differences in the constitution of living bodies and the type of interactions these bodies maintain with the environment (Thompson et al., 1992). The point is that bacteria are sensitive to the environment following their vital norms, just as we are when, after being immersed in the water for some seconds, we look for oxygen in the air to breathe. Sense-making is, therefore, a normative domain of interactions between a living agent and the environment guided by affective bodily processes.

Section 3.4.3 describes the connection between the theory of biological autonomy and nonorthodox accounts of life and evolution. These theories emphasize the central role that organisms' development has in the actual constitution of the living.

3.4.3 Enactive Evolution

In a recent anthology of the philosophy of biology, John Dupré and Daniel Nicholson (2018) provide a manifesto for an ontology of processes in theoretical biology. They claim that:

> [T]he living world is a hierarchy of processes, stabilized and actively maintained at different timescales. We can think of this hierarchy in broadly mereological terms: molecules, cells, organs, organisms, populations, and so on. Although the members of this hierarchy are usually thought of as things, we contend that they are more appropriately understood as processes (p.3).

This ontology of processes diverges from the orthodox definition of life in theoretical biology, that is, as complex physical systems determined by the information supposedly contained in the molecular composition of their genomic structures (2007). The proposed ontology by Dupré and Nicholson instead resonates with Thompson's (2007, p. 166) affirmation that "we living organisms are historical and developmental beings."

From this perspective, the key to understanding the nonorthodox conception of life of the enactive approach is to see it as temporally open-ended. Living systems are oriented toward the future by an intrinsic teleology, looking forward to staying alive. The concrete attainments of this teleology (i.e., the requirements for the self-preservation of organisms) are nonetheless historically determined. That is, the structure of living bodies and their biological needs carry on living systems' phylogenetic and ontogenetic past. This past, nonetheless, does not entirely determine the body and behavior of organisms but merely constrains their unfolding in time. Crucially, organisms remain open to the unexpected contingencies of the present. Consequently, living organisms resemble a path laying down in walking (Varela et al., 2016).

The theory of biological autonomy puts the organism at the center of life's processes. Organisms, in this view, are not substantial things but dynamic processes. Autonomous living systems continuously change and adapt their concrete existence to the current circumstances of the environment, constituting the fundamental unit of life and arguably also of evolution (Thompson, 2007). For the theory of biological autonomy, autopoiesis and autonomy are processes that logically and chronologically precede evolution. For one simple reason, evolution entails reproduction, and for there to be reproduction, the reproduced entity must first exist.

Importantly, reproduction is not replication. Organisms do not simply engender copies of themselves. Biological evolution requires change and variation at the ontogenetic temporal and phylogenetic scales. In this regard, newborn organisms retain *some* of their progenitors' characteristics and diverge from them.

For the orthodox view, variation among different generations of organisms is due to random variations in their genetic material. Some variations are successful in this so-called lottery of life and continue their reproduction, while others are not. The success of a lineage depends on the material condition in the environment. Some organisms will fit more adequately than others to environmental circumstances. Development, in this picture, is nothing more than the process of genetic information transcription and the expression of genes in the appearance and behavior we observe in organisms.

Seen as temporal processes, by contrast, organisms constitute themselves in their concrete development by modulating their interactions with the environment. Developmental processes are nonetheless constrained by structures, inherited phylogenetically, that comprise much more than genes. This is the general idea behind the alternative evolutionary theories of developmental systems theory (Griffiths & Gray, 1994) and evolutionary developmental biology or "evo-devo" (Hall, 1992).

For developmental systems theory, the unity of reproduction of biological evolution is not simply genes, nor even the organism as a systemic unity independent of the environment. Instead, this unity encompasses the network of interdependent processes required for the ontogenesis of organisms, that is, a *developmental system* (Oyama, 2000). Likewise, evo-devo focuses on embryogenesis, holding that alterations in development patterns produce significant changes in evolution. Hence, the visible changes in phenotypes and the changes in the reproduction rates of a species are due to changes in developmental processes and not just to random mutations in genes that the environment eventually selects (Hall, 2012).

Thompson (2007) connects the claims of these theories with the theory of the biological autonomy of the enactive approach. He suggests that altogether these ideas can offer an alternative account of evolution that he calls *enactive evolution*. Such a view of evolution, unlike adaptationism, holds that organisms exhibit the robust (resistant to alterations) and flexible (open to change) structures of development that maintain the continuity of the organism's existence and makes possible its reproduction.

Although there have not been significant works about enactive evolution after Thompson, there is increasing support in theoretical biology for ideas such as an ontology of processes (Nicholson & Dupré, 2018), the agential normativity of living beings (Moss & Nicholson, 2012), and the concept of biological autonomy to explain the systemic complexity of multicellular organisms (Moreno & Mossio, 2015). Much work is still needed to make the project of enactive evolution a solid paradigm in theoretical biology. However, the debates are now open, and there is nothing extravagant or purely speculative in the conception of life and evolution pursued by the enactive approach.

3.4.4 Groundless Grounds

The commitment of radical enactivism to orthodox theories of life and evolution is problematic because these theories neglect the agency and normativity that living organisms seem to exhibit. I have also declared in Sects. 3.3.3 and 3.3.4 that radical enactivism cannot distinguish between efficient interactional and genuinely cognitive systems. In this final section, I explain why the enactive approach can overcome these problems.

Agency is a label we use to attribute the capacity of acting according to their own will to certain beings; for instance, when I decide to extend my arm for a handshake, I am the agent of this act. Another person might respond to this gesture with a similar one, or she may refuse it, which means that she is probably also an agent. We do not attribute agency to machines like blenders and toasters because they do not make decisions to accomplish their functions. Instead, we are the ones who make use of artificial devices for our purposes.

Nevertheless, attributing agency to beings becomes more problematic when we look deeper. On the one hand, contemporary artificial intelligence produces machines designed to look like intelligent beings that can exhibit agent-like behaviors. On the other hand, it is also possible that our purposive intentions are irrelevant for acting. Maybe our actions depend on the blind physical processes that sustain our conscious lives, but what we experience as a decision taken by ourselves is also a causal determination of processes in our brains. Thus, a scientific approach to agency needs a more specific definition.

The enactive approach uses an operative definition of agency based on the requirements of a system for being considered an agent. Barandiaran et al. (2009) distinguish three necessary and sufficient conditions for the agency of a system: ***individuality, normativity, and interactional asymmetry.*** Individuality consists of the system's capacity to determine its own organizational realm. Normativity is the characteristic of a system to behave according to rules that produce positive or negative consequences for the system's organization. Finally, interactional asymmetry implies that a system, even if it is in constant interaction with other systems, can regulate and modify the conditions of its interactions.

It is easy to recognize these three features of systemic agency in the descriptions of life given by the enactive approach. First, an autonomous system is a system that defines its individuality by self-distinguishing from the rest of the environment, sometimes by autopoiesis, sometimes simply by an operational closure. Second, the interactions of such an autonomous system with the environment are fundamentally determined by norms grounded on the self-identity and self-preservation of the system. Finally, the interactional asymmetry exhibited by the adaptive behavior of these systems is due to their capacity to modulate (according to the proper norm) agent–environment system interactions. From this stance, life entails agency and cognition too.

Conversely, artificial sensorimotor systems that generate sensorimotor couplings, such as Brook's robots, do not exhibit agency because they do not possess the intrinsic teleology of life (Froese & Ziemke, 2009). They can be dynamically coupled to the environment and modulate this coupling, but the external observer determines the norms of the coupling. Only living organisms determine their norms because it is a matter of existential concern to sustain and modulate their interactions with the environment. *This fact changes everything*. The intelligent actions of life are due to their existential concerns, bodily affects, emotions, and interests, all grounded in their developmental paths. The "intelligent" behavior of artificial systems, by contrast, is determined by their design and creators' purposes. Even if computational systems can be reprogrammed and updated to produce unforeseen outcomes for their human creators, these tasks are meaningless for the machines, and only the eye of a human agent recognizes the "fantastic" results of their "actions." That does not mean it is impossible to create artificial forms of full-fledged intelligence. It just means that artificial systems must embody the same autonomous organization of life to be truly intelligent (Di Paolo, 2009).

In this manner, the enactive approach *does* provide criteria to distinguish between cognitive agents and artificial systems with efficient behavior. Cognitive agents are sentient beings that behave according to their own norms rooted in their biological constitution. The systemic descriptions of living agents also distinguish between mere self-organizing systems and cognitive systems, which are autonomous systems in precarious conditions with adaptive behavior. That is, cognitive systems are, by definition, living systems. This systemic description of life also tells us that living beings are agents. They are beings that actively change their environmental conditions and their behavior, considering their own purposes. Therefore, the strong enactivism project of the enactive approach offers just the sort of definitions that are missing in the weak enactivism project of radical enactivism (see Fig. 3.1).

Defining life as autonomous is still controversial. Such a claim needs more support and connection with nonorthodox theories of evolution to create a solid alternative to mainstream theoretical biology. However, it is also true that conceiving life as autonomous and normative is an idea gaining support in the field (Moss & Nicholson, 2012). For enactive cognition, the conclusion to draw from this is clear: The enactive approach, or any other strong enactivism that puts autonomy and normativity at the basis of life and cognition, offers a more complete and accurate description of cognitive agents and cognition than those approaches taking raw sensorimotor correlations as the causal bases of this phenomenon.

Enactive Cognition

(Shared assumptions)

- Cognition is rooted in cycles of action and perception
- Perception is guided by action (constituted by sensorimotor loops)

Radical enactivism (as weak enactivism)	**The Enactive Approach** (as strong enactivism)
• It is skeptic about the philosophical grounds of *The Embodied Mind*	• Philosophical foundations: Groundless grounds and Fundamental circularity
• Subjective experience appears later in the evolution of cognitive agents	• Subjectivity is an intrinsic aspect of cognition
• It rejects the theory of autonomy • It describes a heteronomous determination of living organisms	• Cognition is grounded on the biological autonomy of living organisms
• Basic cognition is information as co-variation. High-levels of cognition are contentful, but truth conditions are socially determined.	• Cognition is enaction and sense-making
• It focuses on the evolutionary processes as the causal origins of cognition	• It focuses on development aspects of life and cognition
• It has no clear mark of the cognitive	• Life is the mark of the cognitive

Fig. 3.1 The main differences between radical enactivism and the enactive approach

References

Aizawa, K. (2014). The enactivist revolution. *AVANT, 5*(2), 19–42.

Barandiaran, X. E., Di Paolo, E. A., & Rohde, M. (2009). Defining agency: Individuality, normativity, asymmetry, and spatio-temporality in action. *Adaptive Behavior, 17*(5), 367–386.

Barandiaran, X. E., & Moreno, A. (2008). Adaptivity: From metabolism to behavior. *Adaptive Behavior, 16*(5), 325–344.

Barrett, N. F. (2017). The normative turn in enactive theory: An examination of its roots and implications. *Topoi, 36*(3), 431–443.

Beer, R. D. (2000). Dynamical approaches to cognitive science. *Trends in Cognitive Sciences, 4*(3), 91–99.

Bitbol, M., & Luisi, P. L. (2004). Autopoiesis with or without cognition: Defining life at its edge. *Journal of the Royal Society Interface, 1*, 99–107. https://doi.org/10.1098/rsif.2004.0012

Bourgine, P., & Stewart, J. (2004). Autopoiesis and cognition. *Artificial Life, 10*(3), 327–345.

Bunge, M. (2006). *Chasing reality: Strife over realism*. Toronto University Press.

Churchland, P. (1986). *Scientific realism and the plasticity of mind*. Cambridge University Press.

Clark, A. (1997). *Being there: Putting brain, body, and world together again.* MIT Press.
Colombetti, G. (2014). *The feeling body: Affective science meets the enactive mind.* The MIT Press.
Colombetti, G. (2018). Enacting affectivity. In A. Newen, L. De Bruin, & S. Gallagher (Eds.), *The Oxford handbook of 4E cognition* (pp. 571–588). Oxford University Press.
Colombetti, G., & Thompson, E. (2005). Enacting emotional interpretations with feeling. *Behavioral and Brain Sciences, 28*(2), 200–201. https://doi.org/10.1017/S0140525X05280044
Colombetti, G., & Thompson, E. (2008). The feeling body: Toward an enactive approach to emotion. In W. F. O'verton, U. Müller, & J. L. Newman (Eds.), *Developmental perspectives on embodiment and consciousness* (pp. 45–68).
Cosmides, L., & Tooby, J. (1987). From evolution to behavior: Evolutionary psychology as the missing link. In J. Dupré (Ed.), *The latest on the best: Essays on evolution and optimality* (pp. 276–306). The MIT Press.
Darwin, C. (1859). *On the origin of species.* Oxford University Press.
Dawkins, R. (1982). *The extended phenotype: The gene as the unit of selection.* Oxford University Press.
Dawkins, R. (1986). *The Blind Watchmaker.* W.W. Norton.
Degenaar, J., & O'Regan, J. K. (2017). Sensorimotor theory and enactivism. *Topoi, 36*(3), 393–407.
Descartes, R. (2008). *Meditations on first philosophy: With selections from the objections and replies.* (M. Moriarty, Trans.). Oxford: Oxford University Press.
Di Paolo, E. A. (2005). Autopoiesis, adaptivity, teleology, agency. *Phenomenology and the Cognitive Sciences, 4*(4), 429–452. https://doi.org/10.1007/s11097-005-9002-y
Di Paolo, E. A. (2009). Extended life. *Topoi, 28*(1), 9.
Di Paolo, E. A., Buhrmann, T., & Barandiaran, X. E. (2017). *Sensorimotor life: An enactive proposal.* Oxford University Press.
Di Paolo, E. A., Cuffari, E., & De Jaegher, H. (2018). *Linguistic bodies: The continuity between life and language.* MIT Press.
Di Paolo, E. A., & Thompson, E. (2014). The enactive approach. In L. Shapiro (Ed.), *The Routledge handbook of embodied cognition* (pp. 86–96). Routledge.
Dretske, F. (1986). Misrepresentation. In R. J. Bogdan (Ed.), *Belief: Form, content, and function* (pp. 17–36). Clarendon Press.
Dupré, J., & Nicholson, D. J. (2018). A manifesto for a processual philosophy of biology. In *Everything flows: Towards a processual philosophy of biology* (pp. 3–45). : Oxford University Press.
Egbert, M. D., Barandiaran, X. E., & Di Paolo, E. A. (2010). A minimal model of metabolism-based chemotaxis. *PLoS Computational Biology, 6*(12), e1001004. https://doi.org/10.1371/journal.pcbi.1001004

Fodor, J. A. (1990). A theory of content I: The problem. In *A theory of content and other essays*. Cambridge, MA: The MIT Press.

Froese, T., & Di Paolo, E. A. (2009). Sociality and the life–mind continuity thesis. *Phenomenology and the Cognitive Sciences, 8*(4), 439.

Froese, T., & Ziemke, T. (2009). Enactive artificial intelligence: Investigating the systemic organization of life and mind. *Artificial Intelligence, 173*(3), 466–500. https://doi.org/10.1016/j.artint.2008.12.001

Fuchs, T. (2012). The feeling of being alive. In J. Fingerhut & S. Marienberg (Eds.), *Feelings of being alive* (pp. 149–166). Walter de Gruyter.

Fuchs, T. (2018). *Ecology of the brain: the phenomenology and biology of the embodied mind*. Oxford University Press.

Gallagher, S. (1997). Mutual enlightenment: Recent phenomenology in cognitive science. *Journal of Consciousness Studies, 4*(3), 195–214.

Gallagher, S. (2017). *Enactivist interventions: rethinking the mind*. Oxford University Press.

Godfrey-Smith, P. (2006). Mental representation, naturalism, and teleosemantics. In G. McDonald & D. Papineau (Eds.), *Teleosemantics* (pp. 42–68). Clarendon Press.

Godfrey-Smith, P. (2007). Information in biology. In D. L. Hull & M. Ruse (Eds.), *The Cambridge companion to the philosophy of biology*. Cambridge University Press.

Gould, S. J., & Lewontin, R. C. (1979). The spandrels of San Marco and the Panglossian paradigm: A critique of the adaptationist programme. *Proceedings of the Royal Society of London. Series B. Biological Sciences, 205*(1161), 581–598.

Griffiths, P. E., & Gray, R. D. (1994). Developmental systems and evolutionary explanation. *The Journal of Philosophy, 91*(6), 277–304.

Hacking, I. (1983). *Representing and intervening: Introductory topics in the philosophy of natural science*. Cambridge University Press.

Hall, B. K. (1992). *Evolutionary developmental biology*. Chapman and Hall.

Hall, B. K. (2012). Evolutionary developmental biology (Evo-Devo): Past, present, and future. *Evolution: Education and Outreach, 5*(2), 184–193.

Heidegger, M. (1962). *Being and time*. (J. Macquarrie & E. Robinson, Trans.). : Blackwell Publishers.

Hume, D. (1748). *An enquiry concerning human understanding and other writings*. Oxford University Press.

Husserl, E. (1970). *The crisis of European sciences and transcendental phenomenology: An introduction to phenomenological philosophy*. (D. Carr, Trans.). : Northwestern University Press.

Husserl, E. (1982). *Ideas pertaining to a pure phenomenology and to a phenomenological philosophy I*. (F. Kersten, Trans.) (Vol. II). Dordrecht, NL: Kluwer Academic Publishers.

Hutto, D. D., & Myin, E. (2013). *Radicalizing enactivism: Basic minds without content*. MIT Press.
Hutto, D. D., & Myin, E. (2017). *Evolving enactivism: Basic minds meet content*. MIT press.
Hutto, D. D., & Satne, G. (2015). The natural origins of content. *Philosophia, 43*(3), 521–536.
Jackendoff, R. (1987). *Consciousness and the computational mind*. MIT Press.
Jonas, H. (1966). *The phenomenon of life: Toward a philosophical biology*. Northwestern University Press.
Kant, I. (1998). In P. Guyer & A. W. Wood (Eds.), *Critique of pure reason*. Cambridge University Press.
Lewens, T. (2007). Adaptation. In D. L. Hull & M. Ruse (Eds.), *The Cambridge companion to the philosophy of biology* (pp. 1–21). Cambridge University Press.
Maturana, H., & Varela, F. J. (1980). *Autopoiesis and cognition: The realization of the living*. D. Reidel Pub. Co.
Merleau-Ponty, M. (1964). *Le visible et l'invisible: suivi de notes de travail. par Maurice Merleau-Pont. Texte établi par Claude Lefort*. Gallimard.
Merleau-Ponty, M. (2012). *Phenomenology of perception*. (D. Landes, Trans.). New York, NY: Routledge.
Mill, J. S. (2012). *A system of logic ratiocinative and inductive: Being a connected view of the principles of evidence and the methods of scientific investigation (Books I–II)*. Cambridge University Press.
Millikan, R. G. (1984). *Language, thought, and other biological categories*. MIT Press.
Moreno, A., & Mossio, M. (2015). *Biological autonomy: A philosophical and theoretical enquiry*. Springer.
Moss, L., & Nicholson, D. J. (2012). On nature and normativity: Normativity, teleology, and mechanism in biological explanation. *Studies in History and Philosophy of Science Part C: Studies in History and Philosophy of Biological and Biomedical Sciences, 43*(1), 88–91. https://doi.org/10.1016/j.shpsc.2011.05.007
Myin, E., & Degenaar, J. (2014). Enactive vision. In L. Shapiro (Ed.), *The Routledge handbook of embodied cognition* (pp. 90–98). Routledge.
Nicholson, D. J., & Dupré, J. (2018). *Everything flows: Towards a processual philosophy of biology*. Oxford University Press.
Noë, A. (2004). *Action in perception*. MIT Press.
O'Regan, J. K., & Noë, A. (2001). A sensorimotor account of vision and visual consciousness. *Behavioral and Brain Sciences, 24*(5), 939–973.
Oyama, S. (2000). *The ontogeny of information: Developmental systems and evolution* (2nd ed.). Duke University Press.
Paley, W. (2009). *Natural theology: Or, evidences of the existence and attributes of the deity, collected from the appearances of nature*. Cambridge University Press.

Rietveld, E., Denys, D., & Van Westen, M. (2018). Ecological-enactive cognition as engaging with a field of relevant affordances. In *The Oxford handbook of 4E cognition* (p. 41). : Oxford University Press.
Rowlands, M. (2010). *The new science of the mind: From extended mind to embodied phenomenology*. MIT Press.
Thelen, E. (1995). Time-scale dynamics and the development of an embodied cognition. In R. F. Port & T. van Gelder (Eds.), *Mind as motion: Explorations in the dynamics of cognition* (pp. 69–100). MIT Press.
Thompson, E. (2007). *Mind in life: Biology, phenomenology, and the sciences of mind*. Harvard University Press.
Thompson, E., Palacios, A., & Varela, F. J. (1992). Ways of coloring: Comparative color vision as a case study for cognitive science. *Vision and mind: Selected readings in the philosophy of perception*, 351–418.
van Gelder, T. (1997). Dynamics and cognition. In J. Haugeland (Ed.), *Mind design II: Philosophy, psychology, artificial intelligence* (pp. 421–450). MIT Press.
Varela, F. J. (2000). *El fenómeno de la vida*. Dolmen Ediciones.
Varela, F. J., Thompson, E., & Rosch, E. (2016). *The embodied mind: Cognitive science and human experience* (2nd ed.). MIT Press.
Weber, A., & Varela, F. J. (2002). Life after Kant: Natural purposes and the autopoietic foundations of biological individuality. *Phenomenology and the Cognitive Sciences, 1*(2), 97–125. https://doi.org/10.1023/a:1020368120174
Zahavi, D. (2003). *Husserl's phenomenology*. Stanford University Press.

CHAPTER 4

Body–World Entanglement: On Sense-Making as Norm Development

The enactive approach defines cognition as a form of sense-making. The typical description of sense-making sees the co-emergence of an agent's self-identity, and a correlated meaningful world or *Umwelt*. In addition, a norm that guides the interactions of the nascent agent-environment system also emerges in the act of sense-making. This picture of sense-making implies what Varela called *a surplus of significance*, because meaning—and normativity—appears as a surplus wholly added by the activity of organisms to a physical domain, which is seen as otherwise void of meaning. I will call this standard description of sense-making *the thesis of significance as a surplus*.

In this chapter, I begin by showing that the thesis of significance as a surplus is responsible for several misunderstandings of the ontological and epistemological commitments of the enactive approach. Although I demonstrate these interpretations are wrong, I will argue that more fundamentally this thesis is problematic because it commits a methodological error that fails to describe sense-making adequately (Sect. 4.1).

When we look more attentively at the agent–environment system interactions, we realize there are always previously established normativities that have been enacted in the past of the system. These previously established normativities enable, constrain, and motivate the enactment of new sense-making norms, along with their correlatives of self-identity and *Umwelt*. In this context, sense-making is not an act where a normative and

M. A. Sepúlveda-Pedro, *Enactive Cognition in Place*, New Directions in Philosophy and Cognitive Science,
https://doi.org/10.1007/978-3-031-20282-7_4

meaningful domain arises from a non-normative and meaningless one. Instead, sense-making consists of *reconfiguring* an already existent normativity in the agent–environment system. I call this alternative description of sense-making *the thesis of norm development* (Sect. 4.2).

In support of this thesis, I show that some enactivist works already point to describing sense-making as a process of norm development. That is the case in some descriptions of adaptivity and chemotaxis in *Escherichia coli* bacteria and, more evidently, in the enactivist interpretation of Piaget's theory of equilibration. However, I will argue that Merleau-Ponty's phenomenology of perception is a better guide to understanding the development of perceptual norms as the reconfiguration of the body–world entanglement (Sect. 4.3). These phenomenological descriptions provide the basis for understanding, in subsequent chapters, sense-making as a process of historical and ecological development of the agent–environment system.

4.1 The Thesis of Significance as a Surplus

In the early days of the enactive approach, Varela described an essential difference between the concepts of 'environment' and 'world,' which he called *the surplus of significance* (Varela, 1991, p. 179). This surplus of significance refers to the value that physicochemical events and objects in the surroundings of organisms acquire for them, given their concrete bodily constitution and needs. For instance, the chemical compound 'glucose' has the significance of a 'nutrient' for *E. Coli* bacteria that use this compound for their metabolic functions. Consequently, Varela distinguishes sharply between 'environment,' which stands for the physical surroundings of an organism, and 'world,' which corresponds to the meaningful aspects of these surroundings (those that trigger actions from this organism).

About a decade later, Thompson (2004, p. 386) claimed that living *is* sense-making, which entails that being alive is a continuous process of making sense of the surrounding world according to the constant constitution of the self-identity of organisms as autonomous systems. Sense-making is indeed the co-emergence of (bodily) self and (meaningful) world at once (Thompson, 2007).

However, Thompson (2007, p. 147) also described sense-making as the act of changing the "physicochemical world" into an "environment of significance and valence." For this reason, I will call this traditional way of describing sense-making as *the thesis of significance as a surplus.*

This thesis has been a recurrent source of criticism from the supporters of other forms of world-involving cognition. The problem they see is that sense-making seems to imply the sort of dualisms that characterizes brain-centered cognitive science. Hence, some aspects of the enactive approach could contradict the aims of world-involving cognition.

While I recognize the thesis of significance as a surplus is a problematic description of sense-making, I disagree with the usual arguments against this thesis. In what follows, I clarify the confusing aspects of this typical definition of sense-making in three different stages. In the first stage (Sect. 4.1.1), I explain the meaning of the twofold reference of terms like 'environment' and 'world' within the literature of the enactive approach. These clarifications help us see that the twofold meanings of the terms 'environment' and 'world' of the enactive approach does not entail any of the traditional subject–object or internal–external dualisms. In the second stage of the analysis (Sect. 4.1.2), I argue that the twofold reference of the terms' environment' and 'world' implies a phenomenological dualism that describes a dual aspect of the environment. This dualism does not contradict the purposes of world-involving cognition, though. Finally, in the third stage (Sect. 4.1.3), I claim that the real problem of the thesis of significance as a surplus is its inconsistency with the phenomenological dualism and the philosophical principles of the enactive approach. Later, I will show that looking more carefully at the processes of life and cognition reveals that sense-making consists of norm development.

4.1.1 Stage One: The Environment as Umwelt *and as* Umgebung

Scholars of the enactive approach tend to use the words' environment' and 'world' in two different manners. Sometimes these terms designate the subject-relative world that is significant for the activities of living organisms. At other times, the same words name the subject-independent world described by sciences like physics and chemistry. This twofold semantics causes unnecessary ambiguities.

For this reason, I will import the terminology of the Estonian etiologist Jakob von Uexküll (1934) and name the subject-relative world *Umwelt* and the subject-independent surroundings of organisms *Umgebung*. Given the similarities between the claims of Uexküll and those of the enactive approach (Thompson, 2007), this terminology helps us avoid the semantic ambiguities of this latter approach and improve our understanding of the thesis of significance as a surplus.

In his most paradigmatic example of an *Umwelt*, Uexküll describes three essential features significant for the vital cycle of a tick: (1) an olfactory cue from butyric acid in a mammal's sweat, (2) a tactile cue from a mammal's hair, and (3) the temperature of a mammal's body. For ticks, these three specific aspects of a mammal's body are relevant for getting the blood they need to reproduce (Von Uexküll, 1934, p. 45). The relevance of these aspects depends on the constitution of the tick's sensorimotor system and the vital norms that motivate its sensorimotor behavior. This happens in what Uexküll termed a *functional cycle*. Thick's *Umgebung*, by contrast, comprises all those elements of the environment that we can observe, from our own perspective, in the surroundings of ticks, including those irrelevant to the three functional cycles of these organisms.

We find the origin of the enactive conception of the *Umwelt* in *The Embodied Mind*. Varela and colleagues stated that autonomous systems, while maintaining a structural coupling with the environment, specify *a domain of significance* (Varela et al., 2016, p. 155). This is illustrated by the example of a virtual autonomous system (Bittorio), which, after a period of interactions with its virtual surroundings, self-organizes and begins to select from its random milieu (*Umgebung*) those aspects relevant to its autonomous constitution (*Umwelt*). Later, Varela et al. (2016, p. 181) claim that something similar occurs with living organisms that enact different worlds of color thanks to differences in their sensorimotor systems, biological functions, and environmental conditions.

In his book *Mind in Life*, Thompson made similar descriptions of the meaningful world of *living* sensorimotor agents, saying that "In the case of animal life, *the environment emerges as a sensorimotor world* through the actualization of the organism as a sensorimotor being" (Thompson, 2007, p. 59, my emphasis). In the same work, Thompson makes explicit the parallels between the enactment of a world of significance by an autonomous system and Uexküll's notion of *Umwelt*: "Sense-making changes the physicochemical world into an environment of significance and valence, creating an *Umwelt* for the system" (Thompson, 2007, p. 147).[1] Thompson or any other enactivist never uses the term *Umgebung*. However, for practical purposes, I will use this German word to designate the "physicochemical world" that Thompson mentions in the last citation.

Although supporters of the enactive approach describe, much like Uexküll, two different types of worlds or environments that surround living organisms, they do not necessarily endorse a traditional dualism or posit a radical separation between an epistemic subject and its objective world. De Jesus (2018) thinks the opposite, however. He claims that the enactive approach holds an *epistemological perspectivism* that posits the existence of two kinds of "worlds": the worlds enacted by a living organism (a subjective world), and the world as such (an objective world), over which the organism takes a perspective.

De Jesus seems to misunderstand the claims of the enactive approach. The scholars of this approach do not describe the disclosure of the world by a living body in epistemological terms. The relation between an organism and the environment is fundamentally based on the bodily affectivity of the living organism. This affectivity is constituted and developed in the concrete existence of the organism and its interactions with the environment (Colombetti, 2014). Consequently, there is no epistemological perspective on the world; instead, there is a particular way to make sense of it.

Another related misinterpretation of the thesis of significance as a surplus is the attribution of a "bodily internalism" to the enactive approach. Brain-centered cognitive science is often deemed a cognitive internalism because all the relevant causal processes of cognition are said to happen within the spatial boundaries of cognitive agents or, more specifically, in the head. Similarly, for Wheeler (2010), the theory of biological autonomy entails a new form of internalism since it encloses the relevant causal processes of cognition within the spatial boundaries of the biological body. From this standpoint, the environment may cause perturbations in the living body, but the world remains external and alien to organisms, only providing inputs for their bodily organization.

Wheeler's assertions are nonetheless unjustified for at least two reasons. In the first place, the living body as an autonomous system is, from the beginning, constituted by the agent-environment coupling. The sensorimotor domain of relevance (*Umwelt*) is enacted by sense-making. At the same time, the operational closure or self-making (the other side of sense-making) defines the bodily self of the agent. Neither the body nor the environment is pre-specified in advance of the vital unfolding of the living body. Indeed, thanks to a capacity for adaptivity, living organisms constantly shape and reconfigure their sense-making of the world.

Wheeler's assertions are also unjustified because they fail to appreciate the capacity of the living body to reconfigure its constitution and reshape its boundaries, incorporating objects and processes that were previously only part of the environment (Thompson & Stapleton, 2009). *The boundaries between the body and the world are thus plastic and not predefined* a priori (i.e., independently of their concrete interactions). Beyond otherwise significant differences, the enactive approach holds a perspective similar to externalist accounts of cognition, such as the extended functionalism supported by Wheeler. Indeed, positions like internalism and externalism do not apply to the enactive approach. These two positions typically assume the existence of prespecified boundaries between agents and their environments, but the enactive approach does not (Di Paolo, 2009).

Therefore, the twofold semantics of the term 'environment' for the enactive approach is neither the manifestation of a traditional subject–object epistemic dualism nor the internalism–externalism dichotomy that characterizes brain-centered cognitive science. However, explaining the sort of duality evoked using the twin-terms *Umwelt-Umgebung is still necessary.* In the next section, I argue that the constant shift between the semantic reference of the environment as *Umwelt* and as *Umgebung* is motivated by *a duality of phenomenological aspects of the environment that mirrors the phenomenological duality of the body.*

4.1.2 *Stage Two: The Dual Aspect of the Environment*

From a phenomenological perspective, the environment can be experienced both subjectively *and* objectively, like the body itself. This subsection argues that the enactive approach implicitly endorses the same view.

Following Husserl's (1989) phenomenological descriptions, Fuchs (2018) claims that the body has a dual aspect: It can be conceived either as a lived subjective body (*Leib*) or as a physical or objective body (*Körper*) (see also Legrand, 2006; Thompson, 2005). The lived body refers to our concrete bodily existence, making our world experience possible. My hand, for instance, has a sensitivity of touch and a particular way of articulating its movements to touch and grasp my cup of tea. My experience of the cup, explicitly lived or not, is possible thanks to the capacity of my hand to make sense of the cup by being able to grasp it (Gallagher, 2005). Understood in this way, my hand is part of my lived body.

I can also shift perspective and see my hand with a certain detachment, as being before me or as an object that rests between the cup and my computer. As such, my hand is considered an object that shares the same 'spatial' domain as other objects and is subject to the same kind of 'objective' relations.

Although the body appears to our lived experience both as a subject and as an object, there are not two bodies. The distinction between the two is phenomenological, not ontological. My body has a dual phenomenological aspect (Fuchs, 2018): It can be tacitly experienced as a body (*Leib*) that constitutes my experience of the world, or it can be experienced as an object (*Körper*) that is part of the world I inhabit.

We can find the same dual aspect in the environment. There is, for instance, something like a lived environment, which we understand as the domain we inhabit and with which we are familiar. The lived world of perception primarily consists of what Merleau-Ponty (2012) called motor significations, but that we know more commonly, after Gibson (2015), as affordances. *Affordances are those aspects of the environment that correlate to the motor intentions of what phenomenologists call the lived body.* The cup affords grasp-ability to my hand, the floor offers walkability to my legs, and the chair affords sit-ability to my whole body. Heidegger (1962) called these aspects of the environment *ready-to-hand*. Nailing the wood, an experienced carpenter can sustain a continuous flow between her body, the hammer, and the wood. Dreyfus (1991) named this flow 'skillful coping.' In skillful coping, the perceiver does not need to reflect on her actions or the things her actions are directed to, because there is already at play an attunement between the whole body–world system (Dreyfus & Kelly, 2007). When something goes wrong, and there is a need for significant adjustments in the performance of an action, the carpenter will direct his attention to the environment and its specific aspects that we call objects (Heidegger, 1962).

The objective environment, by contrast, is the space around us, full of objects that exist as such, as if they were independent of our interests and concerns. Heidegger (1962) called this aspect of the environment *present-at-hand*. Perceiving the environment, in this mode of experience, involves more a theoretical than a practical attitude on the part of the perceiver. I can look through the window and see houses, trees, birds, people, the blue sky, and so forth. I can explicitly refer to these aspects of the environment and share my experience of them with others thanks to the acquisition of language. These objects are, therefore, a common place of reference or meanings shared with others.

This dual aspect of experience explains the twofold sense of the terms' environment' and' world' within the enactive approach. The *Umwelt* of the tick is what we think resembles our lived environment. The environment appears before the tick as a dwelling place because it accomplishes there its vital activities, fulfilling its vital norms. When the circumstances are suitable, the tick acts, and, maintaining the flow properly through its three functional cycles, it can reach something like the skillful coping described by Dreyfus. Skillful coping does not involve reflection or other complex cognitive functions that ticks do not possess. Skillful coping can be understood as the proper equilibrium of the agent–environment system (Bruineberg & Rietveld, 2014).[2] All organisms accomplish this equilibrium that satisfies their biological norms. If things go wrong, organisms can modify their behavior (adaptivity) without taking notice of the objects involved in their bodily actions, just as we do daily. Therefore, we can say that the enactive and Uexküll's *Umwelt* of any organism is like the lived environment of ourselves described by phenomenologists (Thompson, 2007).

In any case, we cannot ultimately reduce the lived environment of a human being to an animal-like *Umwelt*. We can detach ourselves and the particular aspects of the environment (objects) we perceive from their immediate perceptual context (Merleau-Ponty, 1942). We can situate these objects in abstract contexts or place them in symbolic horizons that pertain to the shared world of our cultural environments. We also have what Husserl (1970) called a *lifeworld*. In the cultural horizons of a lifeworld, objects can be manipulated abstractly or concretely to accomplish new bodily and cultural practices. Science is one of these practices. The objects we describe in a scientific horizon pertain to a new dimension of meaning, but one that nonetheless refers to the same environment we intend in the unreflective dimension of our perceptual experiences (Thompson, 2016).

Unfortunately, scholars of the enactive approach have never been clear enough about the phenomenological dualism of the environment I have just described. Hence, it is comprehensible the confusion of their critics. The real problems of the thesis of significance as a surplus lie elsewhere, though.

Section 4.1.3 argues that the thesis of significance as a surplus starts with an inappropriate methodology of the enactive approach. This methodology is confusing about the relationship between scientific claims and phenomenological descriptions. It guides us to erroneously think that the contents (scientific claims) of one mode of experience (the intersubjective) are the original foundations of the other mode of experience (subjectivity). Ultimately, this methodological error leads the enactive approach to misdescribe sense-making.

4.1.3 Stage Three: Mutual Enlightenment

Approaching cognition from a dual perspective improves our understanding of this phenomenon both from the scientific and the phenomenological point of view. However, we must be careful in our methodology to avoid dualistic explanations that can miss the ***in-between*** nature of the ***body–world entanglement*** at the heart of the enactive approach. Here is where the real problem of the thesis of significance as a surplus starts.

To recall: The philosophical foundations of the enactive approach entail the thesis of embodied subjectivity, which recognizes subjectivity as an intrinsic aspect of cognition. One of the implications of this thesis is that cognitive scientists need to work under a fundamental circularity. This reminder precludes us from setting the absolute grounds of cognition either in the mind or in the world (see Sect. 3.1), which in turn requires maintaining a constant dialogue between the first- and third-person perspectives of our study of cognition, that is, between science and phenomenology (Varela et al., 2016).

As we have seen in Chap. 3, the third-person perspective of science does not involve, for the enactive approach, knowledge of a subject-independent world but a new form of experience rooted in the intersubjective and enculturated lives of human beings (Thompson, 2016). Describing sense-making as transforming an *Umgebung* into an *Umwelt*, however, commits the error of grounding one mode of experience (the subjective mode) on the experiential contents given in the other mode of experience (the intersubjective, enculturated mode). Therefore, the thesis of significance as a surplus suggests that the absolute grounds of cognition—including its subjective and meaningful dimensions—can be found in the physical world described by science, falling into contradiction with the middle way of the thesis of embodied subjectivity.

We could argue that our scientific, intersubjective mode of experience constitutes a more finely grained description than the purely subjective or ego-oriented mode of experience. For this reason, such an intersubjective account plays a more fundamental role in our understanding of cognitive phenomena. This hierarchical superiority of the intersubjective mode, however, should be understood as only practical and not theoretical. That is, insofar as science is more beneficial for our understanding of cognition (for the practical purposes we have), it could have a priority over phenomenology. Generally, however, the priority of science over phenomenology has led us to misdescribe cognitive phenomena and has produced problematic explanations. This is precisely what Varela et al. (2016) argued in their criticisms of brain-centered cognitive science. However, a similar mistake happens to many enactivists in their classical descriptions of sense-making. I will return to this problem in the following section. Let me first clarify the sort of relation I find more suitable for a cooperation between the enactive approach and phenomenology.

From a methodological perspective, the relation between phenomenology and enactive cognitive science is complicated. There have been efforts to ground scientific knowledge on the transcendental dimension of experience and to base phenomenological claims on our scientific knowledge. The most accepted proposals for the relationship between science and phenomenology in enactive cognition are those that do not intend to ground one sort of knowledge on another but speak, like Gallagher (1997) exceptionally does, of a *mutual enlightenment*.

The mutual enlightenment of cognitive science and phenomenology argues that phenomenological descriptions can constrain and motivate causal explanations, while causal explanations can constrain and motivate phenomenological descriptions of cognitive phenomena. Hence, cognitive scientists should not pretend to find the ultimate roots of one mode of experience in the other but sustain a dialogue between the two, thus maintaining the "groundless grounds" of enactive cognition.

From this standpoint, the gap between the two modes of experience is not a gap that needs to be closed as the so-called explanatory gap of the mind–body problem might. From both the enactivist and the phenomenological perspectives, the gap implied by the dual aspect of the body and the environment has the positive function of forcing us to adjust and attune our scientific and phenomenological claims via the mutual enlightenment.

Following the spirit of the mutual enlightenment, phenomenology should help the enactivist quest for understanding processes of life and cognition. The thesis of significance as a surplus misdescribes sense-making as an activity in which living agents are decoupled from their own ecological and developmental history. However, we can overcome this problem by looking more deeply at the descriptions of phenomenology. Let's consider before how some works of the enactive approach already point to a different way to conceive sense-making.

4.2 The Thesis of Norm-Development

The co-emergence of a self and an *Umwelt* in sense-making implies the enactment of a new norm that guides the interactions of the agent–environment system. The environment is perceived as meaningful insofar as it is relevant to the agent's actions. The agent, by contrast, acquires an identity that distinguishes itself from the rest of the environment. The emergence of a self, an *Umwelt*, and a sense-making norm, however, does not happen in a raw material landscape where emergent processes bring meaning and normativity to nature. At least, this is not a phenomenon we can currently observe in nature.

In processes of life and cognition, we witness the enactment of norms that reconfigure the preexistent normative space that already guides the activity of an agent–environment system. We cannot read if we have not previously acquired the ability to communicate with others using symbolic language. A toddler cannot learn to walk if she has not previously acquired the ability to stand up. A mammal cannot start breathing without the previous ontogenesis of its lungs. Every time we learn a new skill and a new way to make sense of the world, there is a preexistent normative domain that enables and constrains the emergence of a new sense-making norm.

This is not an aspect irrelevant to the study of processes of life and cognition from the perspective of the enactive approach. The norm already at work in the interactions of an agent–environment system shapes sense-making and its possibilities for development. Consequently, I will argue that the entanglement of the body and the world (the already existent normative field between an agent and the environment) is the condition of possibility for sense-making, and sense-making is the continuous development of the norms already at play in this entanglement. I will call this description of sense-making *the thesis of norm development.*

I defend this thesis in the rest of this chapter in three ways. First, I will argue that bacteria chemotaxis and the metabolic needs motivating this chemotaxis imply a body–world entanglement and not a body–world dichotomy, as the theory of significance as a surplus suggests (Sect. 4.2.1). Second, I will show that the account of sensorimotor habits of the enactive approach implies norm development in their dynamical account of Piaget's theory of equilibration (Sect. 4.2.2). Finally, I will argue that, from a phenomenological perspective, the body–world entanglement is the condition possibility for perception and that the similarities between Merleau-Ponty's normative account of perception and the notion of sense-making further support the thesis of norm development (Sect. 4.3).

4.2.1 *The Body–World Entanglement of Living Organisms*

According to the classical descriptions of sense-making, in the paradigmatic example of *E. coli* bacteria moving towards a medium with a high concentration of glucose, bacteria and their surroundings are seen as two separate entities that become linked thanks to the bio-affective responsiveness of the bodies of bacteria to the properties of this chemical compound.

Although this bio-affective bond may accurately describe the relational, normative domain of bacterial bodies and the world's physical properties, we must be aware that such a bond is possible only because of a *previous* bond between the body (of bacterium) and the world. This is not a phenomenon that emerges from or supervenes upon raw physical processes. Contrary to those described in the case of the cellular automata Bittorio, the self-organization processes of living beings do not produce ordered patterns of interaction from a "random milieu" or a chaotic, meaningless world. On the contrary, cognitive and living processes are always *constrained by a previously established normative domain*, *enacted in their own past*, or more precisely, by a set of constraints established in the history of interactions of the living system and the environment.

Chemotaxis is a basic sensorimotor behavior that allows bacteria to swim in a medium, either in a straight line or randomly, due to the articulated movements of their flagella and the sensitivity of these bacteria to the chemical composition of their mediums (Barham, 2012). These two movements are also referred to as *running* and *tumbling*, respectively. E. coli usually run, but when they find locations with high glucose concentrations, they begin tumbling to remain in areas that provide a metabolic benefit. Traditional explanations of chemotaxis say that this happens because E. coli have a sensorimotor pathway that responds mechanically to the presence of certain chemicals (Adler, 1969). Scholars of the enactive approach, on the other hand, see chemotaxis as a normative-oriented behavior that depends on an organism's metabolism (Egbert et al., 2010). If this is the case, running and tumbling are actions that rely on self-organization processes happening at the interior of the operational closure of a bacterium. This is to say that these behaviors occur under self-organization processes, just as much as they depend on the presence of specific chemical compounds in the environment that shape the behavior and bodily constitution of a bacterium. Attaining equilibrium in a bacterium's body through sensorimotor behavior satisfies a fundamental sensorimotor norm (the bacterium flagella's coordinated action). However, this norm is grounded on the vital metabolic norms that maintain the system's viability. The viability of the system represented by a state of organizational equilibrium of a bacterium is a norm that does not simply emerge from the pre-given composition of the body of this organism. The vital norm is instituted as a result of processes of development that cause the expression of specific genes and not others.

Various chemical compounds can benefit *E. coli*'s metabolism, including glucose, lactose, and sucrose. From the standpoint of the enactive approach, the responsiveness of *E. coli* to one or many of these compounds is not determined by dispositional laws, as some authors suggest (Heras-Escribano et al., 2015). It is established in the developmental path of the bacterium. This is more explicitly and carefully described by less typical descriptions of chemotaxis (Barandiaran & Moreno, 2008).

Barandiaran and Moreno (2008) posit that normativity, at the level of life, entails two different kinds of processes: constructive and interactive. The first set of processes consists of those needed to maintain the autonomous organization of a living organism. They are topologically localized in the boundaries of the organizational closure of the system. The second set comprises the interaction processes between the agent and the environment needed to maintain the system's viability. It is important to note that both groups of processes are necessary to preserve the system's viability, that is, both constructive and interactional processes are constitutive of the vital norms of a living organism. When *E. coli* bacteria find low levels of glucose and higher levels of lactose, the bacteria change their constructive processes (their gene expression) to metabolize lactose instead of glucose, adapting its interactional processes to the current conditions of the environment. The change in the body of the bacteria to metabolize lactose causes the emergence of a different normative domain of interaction. Lactose acquires significance because of the circumstances of the whole agent–environment system; the bacteria's body adapts itself to incorporate this compound. The body changes its gene expression adaptively to metabolize lactose instead of glucose in a process known as the Lac Operon mechanism (Barandiaran & Moreno, 2008). This phenomenon involves the phylogenesis of bacteria (inherited genes), adaptive ontogenesis (the expression of specific genes for lactose metabolism), and the current conditions of the environment (an environment with low levels of glucose and higher levels of lactose.

The enactment of the norm of lactose as valuable for the metabolic needs of E. coli is not the result of the constitution of the autonomous system independent of the immersion of this system in its concrete environment. Since the enactment of this norm implies changes in the body of the living agent or its constructive processes, the adaptive behavior of an agent also entails some sort of *incorporation* of the environment. In this case, enacting a vital norm consists of reconfiguring the whole agent–environment system and not simply of how the agent makes sense of the environment.

Therefore, from an enactive perspective, when we describe the emergence of a vital norm, like *E. coli* bacteria "running towards glucose," we should not state that an organism projects meaning on a raw physical substance. It is more proper to say that an organism *incorporates* an aspect of the environment into its already constituted interactional domain. Incorporation, in this case, means a reorganization of the already existent body–world entanglement of the agent–environment system, the acquisition of a new sense or a new norm for the sort of interactions this system maintains.

The origin of vital norms is not only ontogenetical but also phylogenetical (Zimmer, 2008). The descriptions of the so-called enactive evolution (the connection between the theory of biological autonomy and non-orthodox theories of evolution) should help us explain the temporal connection between different generations of organisms and the processes of autonomy and reproduction (Thompson, 2007). I cannot go further into this question here; to do so would require a careful treatment of genetic and epigenetic processes beyond this book's scope. However, we must bear in mind that the development of any organism is constrained by processes that involve the inheritance of a developmental niche that includes genomic and extra-genomic material (Arnellos, 2018). Section 4.2.2 addresses the second and most important argument for the thesis of norm development, namely, the account of the enactive approach to sensorimotor habits, which is the level where cognition, properly speaking, begins.

4.2.2 Sensorimotor Norms: Sense-Making as Norm Development

The enactive approach to the sensorimotor level of autonomy and sense-making is built on the theory of sensorimotor contingencies. This theory describes perception as the capacity of an individual to master lawful correlations of sensory inputs and motor actions (O'Regan & Noë, 2001). However, in a careful analysis of this theory, Buhrmann et al. (2013)

realized that a more precise definition of sensorimotor contingencies was needed, as well as an adequate account of the mastery or 'know-how' of these contingencies. The enactive approach proposed to create an operational definition of sensorimotor contingencies based on the conceptual framework of dynamical systems theory. This redefinition of sensorimotor contingencies allowed the enactive approach to recognize the same organizational features of life at play in the constitution of sensorimotor habits.

The classification of the operational framework of the enactive approach distinguishes four types of sensorimotor contingencies: ***sensorimotor environment, sensorimotor habitat, sensorimotor coordination, and sensorimotor schemes*** (Di Paolo et al., 2017). A sensorimotor environment describes the lawful correlations between sensorial stimuli and movement, such as changes in the elliptic form of a circular object when we take a particular view of it (Noë, 2004). The sensorimotor environment ignores, for convenience, certain changes that occur in the specific movement of a body. A sensorimotor habitat, by contrast, denotes the changes occurring in the body and the brain that determine the enactment of a particular sensorimotor correlation, such as the specific motor actions needed to recognize the softness of a sponge. Sensorimotor coordination describes how multiple sensorimotor correlations interact to accomplish a particular bodily action, like the bundle of motor movements needed for grasping a cup of coffee. Finally, a sensorimotor scheme is a sensorimotor coordination with a specific normative domain. That is, sensorimotor correlations coordinate in such a way that they deliver good or bad outcomes for accomplishing a bodily action.

It is important to emphasize that the first three types of sensorimotor contingencies are abstract ways of understanding the possible correlation between motor actions and perceptual feedback. In the notion of sensorimotor schemes, we begin to uncover the actual phenomenon of action and perception simply because cognition, as sense-making, is necessarily normative according to the theory of biological autonomy. Thus, the most critical task is to discover where the normativity of sensorimotor schemes comes from if it does not wholly rest on the biological level of autonomy.

Sensorimotor habits constitute the normative domain of sensorimotor actions. Bodily habits are generally composed of more than one sensorimotor scheme. Preparing coffee in the morning, for instance, involves a temporal sequence of different sensorimotor schemes. I look for the bag of coffee and filters in the cupboard, then I put the filter in the machine and add the coffee. After this, I put the water in the machine. Finally, I push the button to turn on the machine. Other activities like swimming, by contrast, require me to coordinate multiple sensorimotor schemes synchronically, such as kicking, making stroking movements with my arms, and breathing.

The recurrence of any successful sequence or a synchronic bundle of sensorimotor schemes produces a tendency to repeat the same dynamic network in similar situations. Di Paolo et al. (2017) have argued that this process is akin to how neurons tend to be activated together when the same input is present.

The interdependency of multiple sensorimotor schemes to preserve sensorimotor habits suggests they have an operational closure (Di Paolo et al., 2017). Hence, sensorimotor habits acquire autonomy because multiple sensorimotor schemes are interdependent diachronically or/and synchronically for the suitable accomplishment of bodily action.

However, the dynamic nature of the world, environmental circumstances, practical contexts, and the physical conditions of the body are in constant change. Hence, the autonomous organization of habits is subject to precariousness, that is, the operational closure of networks of sensorimotor schemes can be lost. Sensorimotor agents can conveniently handle the world's contingencies and regulate their interactions to sustain the autonomous organization of sensorimotor habits. Thus, interactional asymmetry is also present at this level (Di Paolo et al., 2017). Habits acquire their own normativity, which depends on sustaining the same sensorimotor networks that constitute a sensorimotor habit.

Bodily habits are usually part of wider contexts of action. I make coffee for breakfast, swim in a public pool, or cook dinner for my family. The broader networks of habits that constitute a context of action are called a *microworld* (Varela, 1999). In these microworlds, the dynamic coupling of multiple sensorimotor habits becomes mutually sustained, as sequences or synchronic bundles present for accomplishing a series of actions. That is, there is a fractal reproduction of the interdependency of sensorimotor networks that goes from sensorimotor coordination to sensorimotor schemes, from sensorimotor schemes to sensorimotor habits, and from sensorimotor habits to microworlds (Di Paolo et al., 2017). There is a mutual reinforcement and a tension between the different fractal levels of sensorimotor autonomy that creates a highly dynamic and stable sensorimotor life.

A dynamical interpretation of Piaget's theory of equilibration illustrates the interplay of stability and change in the sensorimotor life of agents (Di Paolo et al., 2014). Two processes are crucial here: *assimilation* and *accommodation*. Assimilation refers to integrating an environmental aspect into the agent's physiological or cognitive/behavioral structure. When an unexpected contingency appears in sensorimotor actions, an agent–environment sensorimotor system must recognize itself to assimilate this contingency. Therefore, the dynamic coordination of the body and the world must agree with a sensorimotor norm (Di Paolo et al., 2017, p. 84). Accommodation then describes the processes through which an agent modulates its physiological and behavioral structures to facilitate the assimilation of an aspect of the environment that is not yet assimilated. Equilibration is, therefore, the process by which a sensorimotor organization reaches new stability, reducing the tension and the disparity caused by encountering the unexpected.

The process of agent–environment system reconfiguration that characterizes sense-making is the sort of phenomenon that Piaget's theory of equilibration describes. Piaget describes the process of assimilation as incorporating a new environmental aspect into the interactional domain of the agent–environment system. Accommodation, by contrast, describes the adjustments the agent needs to accomplish to incorporate or assimilate the recently encountered aspect of the environment. For Piaget, accommodation seems to consist of the adjustments an agent performs in its own body. In contrast, for the enactive approach, these changes can also occur in the environment (Di Paolo et al., 2014). From a dynamical interpretation of Piaget's theory, given the coupling of the agent and the environment, changes in the parameters of any of the two systems can produce changes in the constraints of the whole coupled agent–environment system.

Perhaps the best example and illustration of incorporation from a Piagetian perspective is one given by Di Paolo et al. (2014) in their dynamical description of how the bodily coordination of three basic sensorimotor schemes (sucking, breathing, and swallowing) needed by newborns to feed from the breast of their mothers eventually lead agents to acquire new sensorimotor schemes like feeding from a bottle of milk. The acquisition of new sensorimotor schemes involves adjusting the already existent sensorimotor coordination of the baby (the already established configuration of the agent–environment system) to enact a new normative domain of sensorimotor interactions. From the perspective of an external observer, there was a bottle of milk before and after the baby could feed from it. However, from the baby's standpoint, there was nothing but a disruption or an alteration of the already established order when the bottle was not like her mother's breast. There was a concurrence of chaos and order at the same moment the baby first encountered the bottle, as well as a non-coincidence between the expectation sedimented in the past (the habit of breastfeeding) and the encountered new situation that diverged from what the agent expected. This encounter between the already established norm (breastfeeding), and the divergence of the new situation from this norm, enables the emergence of a new sense or a new norm (bottle-feeding).

Chemotaxis and sensorimotor skills learning in human babies show that norms of sense-making are not enacted because of an act of co-constitution of a self-identity and a meaningful world from a chaotic and meaningless background. The previous structures of the agent's body determine the emergence of sense-making norms. However, it is still worth asking if sense-making as norm development implies only the normative constraints previously incorporated into the agent's body and if incorporating the environment to the realm of sense and meaning entails the passage of the purely material environment to the agent's *Umwelt*.

If there are previous norms incarnated in an agent's body, there should be a previous *Umwelt*, and if that is the case, then there should also be normative constraints sedimented in the *Umwelt*. These constraints, like those incarnated in the body, would predispose the following acts of sense-making, exposing sense-making as an act of reconfiguring the normative domain of the agent–environment system as a whole. That is what the thesis of norm development intends to show.

In Chap. 5, I more appropriately describe the agent–environment system normative domain's transformation process. We need first to look at the phenomenological basis that can help us understand norm development more clearly. More specifically, Merleau-Ponty's normative account of perception provides us with conceptual tools for understanding this not always evident phenomenon.

4.3 The Phenomenology of Norm Development

As mentioned in Chap. 2, phenomenology has significantly influenced the enactive approach. For example, in *The Embodied Mind*, phenomenological resources helped shape the thesis of embodied subjectivity and opened the way to acknowledge the fundamental circularity of cognitive science. Later, phenomenology provided methodological resources to create experimental frameworks based on the rigorous analysis of subjective experience, as seen in the neurophenomenology project (Lutz &

Thompson, 2003; Varela, 1996). Another important way phenomenology has influenced the enactive approach is by disclosing cognition's bodily and worldly roots from a transcendental perspective (Thompson, 2007).

A transcendental philosophy does not make claims about the causal bases of phenomena but studies the conditions of possibility for phenomena as given in experience. Kant, the first transcendental philosopher, attempted to make the claims of his transcendental philosophy universally valid. Husserl pursued a similar path. However, phenomenology itself can disclose the limitations of phenomenological analyses and oblige us to be more modest about the transcendental character of our claims (Merleau-Ponty, 1964).

Like Merleau-Ponty (1964) did in his latest works, Gallagher (1997) also refers to phenomenology as a methodology that helps us uncover *the invariants of experience*. These invariants seem essential, eidetic, or even structural components of cognitive phenomena. However, we may fail to describe them correctly, both because phenomenology has its methodological limitations and because the analysis of any worldly phenomena is, in the end, inexhaustible (Merleau-Ponty, 2012). Therefore, if we find empirical evidence that contradicts our phenomenological theses, we can adjust and refine them, thanks to scientific knowledge.

The mutual enlightenment between enactive cognition and phenomenology I mentioned before (Sect. 4.1.3) helps us guide our scientific theories phenomenologically. Scientific theories, by contrast, constrain, motivate, and refine our phenomenological theses (Gallagher, 2012). For example, the enactive approach can assist phenomenology in understanding the role of life in the constitution of lived experience, the nature of the living body, and the link between life and the natural world. By contrast, phenomenology can guide and constrain the development of the concepts and theories of the enactive approach. From this perspective, phenomenology can guide us to understand and define better sense-making.

At this point, it should be clear that the enactment of a world of significance, described by the concept of sense-making, does not mean the production of a representational, symbolic, or any other formal world from the raw matter of the physical world. Instead, it means enacting a normative domain of interactions between a living agent and its environment. Norms of interaction are not aleatory nor determined from the outside of autonomous systems; instead, the existential condition of the living body determines the correct or incorrect character of the interaction.

Normativity and meaning, in this context, are two intimately interrelated concepts. The meaning or significance of an aspect of the environment, for instance, is determined by the norm that makes its appearance possible: Glucose 'appears as food' thanks to its usefulness for the metabolic requirements of *E. coli*. However, we must not think that the norm is something over and above the meaning that the norm expresses (Morris, 2018). The norm is not an a priori determination of the body or the environment: The appearance of the 'glucose as nutrient' for *E. coli* expresses at once the metabolic needs of *E. coli* (vital norm), their chemotaxis skills (sensorimotor norm), and the relevant properties of the environment for these activities. As the enactment of significance, sense-making is the enactment of a normative domain or a set of actions and perceptions with a specific orientation for the agent–environment system.

From a phenomenological perspective, bodily action and perception also involve normativity and meaning (Crowell, 2013), although in a different sense. The phenomenological descriptions of Husserl and Merleau-Ponty on this subject have been central to developing the enactive approach. First, however, let me review some of the main ideas of these two phenomenologists about the normativity of perception to substantiate my argument that sense-making is a phenomenon of norm development.

4.3.1 Husserl's Theory of Perception: Temporality and Horizonality

We can trace it back to Husserl's conception of intentionality, the normative or meaningful character of perceptual experience can be, which he developed in opposition to Brentano's. For Husserl, the relation between perceptual consciousness and its object is intrinsically normative and meaningful (Crowell, 2013). The normativity of perception in Husserl's phenomenology is best understood with regard to the *structure of fulfillment* (Husserl, 1900).

In the case of perception, and contrary to other modes of consciousness, the structure of fulfillment is not derived from any other content of consciousness (e.g., a perception, a memory, a fantasy, etc.) but from the presence of the object itself (*in the flesh*). This presence is nonetheless always partial or perspectival; as such, it generates what is best known as *the problem of perceptual presence* (Noë, 2004). Simply put, the problem is that I only have direct (sensorial) contact with one object profile (e.g., I only see the upper part of the tree from the view of my window). Husserl's idea is that I see the object itself *through* this profile. This is possible because I do not experience the presence of this profile alone. While I only see the upper part of the tree, I can still experience the presence of the tree as a whole. Husserl explains the tension between these two points of view by having recourse to the structure of fulfillment: the absent (or co-given) profiles of the object nonetheless pertain to its meaningful structure; as such, they demand to be fulfilled. Therefore, perception is normative because perceived objects are lived as normal insofar as they fulfill the intentions of consciousness, for example, fulfilling my expectations to find the lower part of the three when I lean out of the window (cf. Doyon, 2021). The question, at this point, is how, from a phenomenological perspective, we can define and explain the origin of the norm that establishes the criterion of fulfillment.

The temporal constitution of consciousness is the key to understanding the interplay of presence and absence exhibited by perceptual phenomena. Husserl's (1991) analysis of time-consciousness departs from our perceptual experience of temporally extended objects such as musical melodies. Lived melodies are meaningful unities that make sense only insofar as they carry on their own past and look forward to a continuity of the present in the near future. A single tone without any previous or subsequent tones is not a melody. It is a mere musical note. Husserl realizes that the experience of temporal objects involves a constitutive "synthesis" of three different elements: (i) what is given at the present moment (*primal impression*), (ii) the immediate past moment (*retention*), and (iii) what is about to happen (*protention*) (Husserl, 1991, §16). This synthesis does not simply involve juxtaposing three similar moments, nor are they superimposed objects. Instead, the synthesis consists of an intertwining of different intentions, involving a normative interplay between presence and absence that gives us the experience of a transition of various aspects of the same object.

Husserl elucidates the paradoxical situation of identity in difference and presence in absence in his postulation of a *double dimension of intentionality*, where one dimension intends the transitional phases (retentions and protentions). In contrast, the other dimension intends the actual living present. Thus, we experience the extended temporality of objects by articulating the retentional and protentional phases, while the flow and the actual living present are *simultaneously* intended (Zahavi, 2003, p. 84). The most relevant aspect of this description is that absence or emptiness plays a constitutive role in the experience of temporally extended objects.

Absence and emptiness, nonetheless, do not mean pure nothingness since they entail the presence of something that is missing and expected. Hence, to acquire a normative status (meaning), these objects must be constituted not only by presence but also by a lack, a hollow, or a constitutive absence.

This description of the tripartite structure of lived temporality is not exclusive to temporally extended objects such as melodies. Husserl's late works also considered the implication of this structure for the perceptual experience of *all* objects. Since the presence of any object involves the synthesis of multiple profiles into a single unity (putting aside the controversial concept of noema in Husserl's transcendental period), the link between the different profiles of any object is given by the possibilities for bodily action that the present profile motivates (Husserl, 2001).

One of the best means for approaching this later understanding of the relation of presence and absence, or the determinable indeterminacy of perception, is through Husserl's notion of *horizons*. Husserl defines a horizon of experience as follows:

> What is currently perceived, what is determinate (or at least somewhat determinate) and co-present in a more or less clear way is in part pervaded, in part surrounded by a horizon of indeterminate actuality, a horizon of which I am dimly conscious. I am able to direct the illuminating focus of my attention on it with varying success. (Husserl, 1982, §27).

Horizons are thus those parts of appearances that are not explicitly intended but are nevertheless co-present with the explicitly intended objects, such as the back of my computer screen. It is worth noting that horizons are not just marginal aspects of our experiences but also constitutive ones. For example, when I am typing on my laptop, I suddenly turn my head and notice the presence of a cup of coffee. I feel the impulse to grasp the cup to sip from it, but my attention is on my writing. Despite this, I successfully hold the cup and drink some coffee. In this situation, I never doubted the shape of the cup or the movement of the liquid inside when I brought the cup to my mouth. I just took the "right" hand position. My experience of the cup was never permeated by any kind of skepticism about the backside of the cup or the volume of the liquid inside. Yet, none of these visual profiles were currently given. All these co-present aspects are what Husserl would call horizonal. However, their relation to the manifestation of the cup is not only temporal but corporeal because the sense or meaning of the horizonal aspects of the appearance involves certain expectations that are related to my bodily skills and interests:

> Every visual sensation or visual appearance that arises in the visual field, and every tactile appearance that arises in the field of touch, is ordered with respect to consciousness. For each current situation, there is a horizon of further possibilities that are ordered together, creating a horizon of possible series of appearances belonging to the freely possible series of movement. (Husserl, 2001, §3).

Consequently, the spatiotemporal horizons of lived things are *not* abstract, ideal, or representational aspects of things because they cannot be given as positive contents of consciousness. There are at least two reasons for this. First, it is not reflective consciousness (cogito) that intends these horizonal aspects since, at the reflective level, perceptual (ready-made) objects are the intended contents of experience. Here, the horizons are best understood as pre-objective. This means that horizonal aspects of experience are given at the pre-reflective (Merleau-Ponty, 2012) and pre-objective or pre-noetic (Gallagher, 2005) levels of experience. Second, and more importantly, the indeterminacy of perceptual appearances is inherent to the thing itself and not to consciousness because it is the thing itself that "hides" its profiles from my sight. The thing remains open to fulfill or not my protentions and expectations, and it is only thanks to this constitutive non-givenness of the thing that it can appear to me as a thing (Merleau-Ponty, 2012, p. 341).

Nonetheless, the givenness of the thing itself is not referred to by my consciousness as a reflective ego (cogito) but to my body, which makes sense of the situation without my reflective consciousness taking notice of this intentional relation between things and the body. Husserl called this pre-noetic level of intentionality *operational intentionality*. This intentionality describes the directedness of a *pre-reflective ego* (an embodied perspective) toward something that appears, not as an entirely determined object, but as something impregnated by a form of meaning that Husserl and Merleau-Ponty called *sense* (*Sinn*).

Let us go back to the problem of perceptual presence, where the horizons of the thing are related to the motor skills of my body because the unseen profiles of the cup are, for example, senseful aspects for my hand to grasp, for my mouth to sip, and so on. Here, the operative intentionality is *motor intentionality* (Merleau-Ponty, 2012, p. 112). However, my interest in drinking coffee also determines the motor significance of the cup. This signification can change indeed if I shift the practice I'm doing (e.g., I will hold the cup differently if I wash it). Therefore, the "bodily meaning" or the *motor signification* (Merleau-Ponty, 2012, p. 113) of the cup is founded on my body's protentions and expectations of the perceived thing.

There are descriptions of these motor significations in Husserl's theory of perception. He posits that the unseen profiles of visual objects are given by a correlation between kinesthetic sensations and changes in the sensorial data (Doyon, 2018). Nevertheless, Merleau-Ponty made the most systematic phenomenological analysis of the role of bodily self-movement in the constitution of perceptual objects. The following subsection will consider his central claims in this respect.

4.3.2 The Body–World Entanglement

In his *Phenomenology of Perception* (2012), Merleau-Ponty describes the constitutive role of the body for perceptual experience through his notion of the *body schema*. For Merleau-Ponty, the body schema refers to the capacity of the body to self-organize without the intervention of reflective consciousness for accomplishing a practical task (Merleau-Ponty, 2012, p. 103). As a phenomenological term, the body schema is more than this because it also points to the deep entanglement of the body and the world, which we cannot entirely separate, even in reflective analysis. This profound level of the body schema is expressed by Merleau-Ponty in the *Phenomenology* when he claims that "the theory of the body schema is implicitly a theory of perception" (Merleau-Ponty, 2012, p. 213). At this pre-reflective level of the bodily constitution of consciousness, we cannot restrict what is on the side of the body and what is on the side of the world because the bodily ego and the world are not originally independent unities. They are polarities that emerge from the self-movement and development of life and nature.

The notion of the body schema is nevertheless commonly associated with the sensorimotor unity of the bodily-ego pole and the *phenomenological field* with the worldly pole of *the body–world entanglement*. This notion resonates with Heidegger's (1962) notion of being-in-the-world but aims to emphasize the importance of the body for the factual existence of a *Dasein*, something neglected by Heidegger. Merleau-Ponty ambiguously uses the term body schema to describe the ego pole and sometimes refers to the body–world entanglement. For the sake of clarity, I will keep the notion of the body schema to refer to the self-organizing capacity of the body or the unity of the sensorimotor skills of the ego pole. From here, I will use the concept of body–world entanglement to denote the foundational sensorimotor unity of the body and the world.

The notion of the body schema is helpful not only to describe the protentional aspects of spatial objects by claiming that this body schema intends to the motor significations of things, but it also helps us explain the presence of the past in the perceptual disclosure of things. The body schema makes concrete the double dimension of temporality incarnated in a corporeal being that unfolds itself in time, entangled with a world that transcends the (bodily) self of lived experience and co-constitutes the bodily self. Although the body schema is an a priori, or a transcendental condition, for perceptual experience, the body schema is not purely transcendental (Morris, 2004). The constitution of the body schema depends on the concrete temporal unfolding of the living body and its concrete developmental history. Merleau-Ponty describes indeed how the acquisition of motor habits reconfigures the body schema to the point of incorporating artifacts and tools to the biological constitution of the body, thereby generating new motor significations and even a new bodily self (Merleau-Ponty, 2012, p. 143).

Hence, the body schema is an open-ended structure that remains subject to change in light of its encounters with things and others. In the case of humans, many of the habits we acquire, if not all, are mediated by our social interactions. Other people shape how we make sense of things as they guide our attention, constrain our behavior, or affectively motivate our actions (Gallagher, 2005).

Given the open-ended nature of things and the body schema, the intertwining of the body–world entanglement is dynamic. Our analysis of its unfolding represents a new way of approaching the mind and world relationship. From this phenomenological standpoint, the problem is not to find causal relations between the body and the world, seen as separate, but to disentangle the thickness of their intertwining and understand the logic of their temporal unfolding.

4.3.3 Perception, Sense-Making, and Temporality

Merleau-Ponty's account of motor intentionality, along with his descriptions of the kind of norms at play in action and perception, has significant similarities with the descriptions of sense-making provided by the enactive approach. In both cases, the world appears to bodily subjects as already meaningful thanks to their bodily skills. From a third-person perspective, the body schema is the unity of the body that self-organizes according to sensorimotor norms; it is the task at hand that determines the body's articulation and the appearance of the world. From a transcendental perspective, the world appears before consciousness as it affords possibilities for action to the body schema. There is thus a co-determination of the body and the world, as Thompson describes sense-making (Thompson, 2007). However, there is also an irreducible body–world entanglement.

This entanglement is more than a phenomenological co-determination, and it is more than a causal coupling. The co-determination of the body and the environment refers to the process by which a (systemic) self-identity is correlated to the identity of otherness (what falls outside the system but is relevant to the system). For example, the grasping of my hand is co-determined by the graspability of the cup. The causal coupling of the agent-environment system allows the system to co-evolve in multiple ways. The environment constrains the bodily activity of the agent, shaping its constitution and dynamic interaction with this environment. The agent acts upon the environment and changes its conditions, transforming how the environment affects the agent's body. Therefore, there is a mutual co-determination and an intimate causal coupling of the agent–environment system, but the body–world entanglement is deeper than this.

The body–world entanglement is the necessary condition for any norm of sense-making or motor intentionality to appear.[3] As I mentioned, Merleau-Ponty describes this entanglement when defining the body schema. Still, he also referred to this entanglement in its account of temporality, like when he talks about "the past that has never been present" (Merleau-Ponty, 2012, 252).

The nature of the body–world entanglement is essentially temporal because it always entails a past (a previously given set of normativities) that can be actualized in the present, thanks to the open-ended nature of the entanglement. That is, the entanglement is permanently open to its actualization due to the encounter of the unexpected events that characterize the nature of the world (Dastur, 2000).

However, the past is not a determination only due to the open-ended nature of the agent–environment system. The enactment of new norms can also transform the past incarnated in the body and the lived environment. I can, for instance, lose a bodily habit (e.g., smoking) that I have acquired in the past thanks to the acquisition of a new habit (e.g., jogging). The new habit sets provisional or definite new constraints that can impede my body from reenacting the old norm. In this sense, my body acquires a new orientation or sense that changes or reconfigures its own past instead of merely accumulating new norms in addition to the older ones.

Although the body–world entanglement is a transcendental claim (a condition of possibility for motor intentionality), it is neither unfamiliar nor impossible to fit this concept in the framework of the enactive approach. The already given presence of the body–world entanglement is expressed in the existence of previous normativities that constrain and enable new forms of sense, as the dynamical account of Piaget's theory of equilibration illustrates (Di Paolo et al., 2014).

The wrong picture for the enactive approach is to think that sense can originate at some specific point from laws of causality that abstract the organism from the environment or the environment from the organism. We could argue that scientific models of sense-making must carry out this sort of abstraction to clarify certain aspects of this phenomenon and give greater explanatory power to the theories of the enactive approach. However, we should see it as a heuristic resource only. It might apply to cases where our analysis of the cognitive phenomena is very general or when we want to illustrate the differences between a purely physical environment and a cognitive domain of significance. However, suppose we require to be more precise and offer more accurate and concrete models of cognition, as we often need in some scientific areas like psychology and psychiatry (Durt et al., 2017). In that case, we need to stop thinking of sense-making as adding a surplus of significance to the physical environment and start thinking of sense-making as a process of norm development.

The thesis of norm development also implies a revision of our conception of the environment from an enactivist perspective. If agents never interact with a raw physical realm but with an environment that is already loaded by a pre-given set of normative relations, then we need to understand the environment as a field of normative forces and constraints that will be central to shaping the path of living and cognitive agents. Defining the enactive environment is the main subject of Chap. 5.

Notes

1. Uexküll did not go deeper into explaining the origins of an animal's functional cycles. Thus, there lies open the possibility of fitting Uexküll's work into an account of life that is rigorously focused on development, such as the enactive approach is, or into more traditional explanations of life and evolution (see, e.g., Heras-Escribano & de Jesus, 2018). Such an exercise is none-

theless irrelevant to the common definition of the enactive approach, as well as to Uexküll's conception of *Umwelt*.

2. The similarities between skillful action or know-how and processes of sense-making are highlighted by Rolla and Huffermann (2021).
3. This statement does not mean that any new form of sense necessarily arises due to the historical past of the body–world entanglement. It rather means that any new form of sense needs to involve this past. For instance, archeologists and linguists cannot decipher ancient and unknown languages but using linguistic and symbolic schemas they already have at hand. The ancient language is for the scientist a new language, and it remains indecipherable and senseless unless a scientist can "relocate" this language in our current networks of meaning, although enacting this network in a whole new way that partially reconstitutes the ancient language (see, e.g., the case of Olmec writing: del Martínez et al., 2006). Similarly, a brain or bodily injury can abruptly alter the body–world entanglement and affect the continuity of the interactional system (e.g., leading people to lose skills and the correspondent perception of affordances). The transformation of the system to equilibrate the new situation needs to reorganize the system but from the remaining structures of the system. This is to say that in such cases patients need to create new forms of sense or new norms of interaction from the already existent forms of sense (cf. Toro et al., 2020).

References

Adler, J. (1969). Chemoreceptors in bacteria. *Science, 166*(3913), 1588–1597.

Arnellos, A. (2018). From organizations of processes to organisms and other biological individuals. In J. Dupré & D. J. Nicholson (Eds.), *Everything flows: Towards a processual philosophy of biology* (pp. 199–124). Oxford University Press.

Barandiaran, X. E., & Moreno, A. (2008). Adaptivity: From metabolism to behavior. *Adaptive Behavior, 16*(5), 325–344.

Barham, J. (2012). Normativity, agency, and life. *Studies in History and Philosophy of Biological and Biomedical Sciences, 43*, 92–103.

Bruineberg, J., & Rietveld, E. (2014). Self-organization, free energy minimization, and optimal grip on a field of affordances. *Frontiers in Human Neuroscience, 8*, 599. https://doi.org/10.3389/fnhum.2014.00599

Buhrmann, T., Di Paolo, E. A., & Barandiaran, X. (2013). A dynamical systems account of sensorimotor contingencies. *Frontiers in Psychology, 4*, 285.

Colombetti, G. (2014). *The feeling body: Affective science meets the enactive mind*. The MIT Press.

Crowell, S. G. (2013). *Normativity and phenomenology in Husserl and Heidegger*. Cambridge University Press.

Dastur, F. (2000). Phenomenology of the event: Waiting and surprise. *Hypatia, 15*(4), 178–189.
De Jesus, P. (2018). Thinking through enactive agency: Sense-making, bio-semiosis and the ontologies of organismic worlds. *Phenomenology and the Cognitive Sciences, 17*(5), 861–887.
del Martínez, M. C. R., Ceballos, P. O., Coe, M. D., Diehl, R. A., Houston, S. D., Taube, K. A., & Calderón, A. D. (2006). Oldest writing in the New World. *Science, 313*(5793), 1610–1614.
Di Paolo, E. A. (2009). Extended life. *Topoi, 28*(1), 9.
Di Paolo, E. A., Barandiaran, X. E., Beaton, M., & Buhrmann, T. (2014). Learning to perceive in the sensorimotor approach: Piaget's theory of equilibration interpreted dynamically. *Frontiers in Human Neuroscience, 8*, 551.
Di Paolo, E. A., Buhrmann, T., & Barandiaran, X. E. (2017). *Sensorimotor Life: An Enactive Proposal.* Oxford University Press.
Doyon, M. (2018). Husserl on perceptual optimality. *Husserl Studies, 34*(2), 171–189.
Doyon, M. (2021). The normative turn of perceptual intentionality and its metaphysical consequences (or why Husserl was neither a disjunctivist nor a conjunctivist). In H. Jacobs (Ed.), *The Husserlian mind.* Routledge.
Dreyfus, H. L. (1991). *Being-in-the-world: A commentary on Heidegger's being and time, division I.* MIT Press.
Dreyfus, H. L., & Kelly, S. D. (2007). Heterophenomenology: Heavy-handed sleight-of-hand. *Phenomenology and the Cognitive Sciences, 6*(1–2), 45–55.
Durt, C., Fuchs, T., & Tewes, C. (2017). Embodiment, enaction, and culture.
Egbert, M. D., Barandiaran, X. E., & Di Paolo, E. A. (2010). A minimal model of metabolism-based chemotaxis. *PLoS Computational Biology, 6*(12), e1001004. https://doi.org/10.1371/journal.pcbi.1001004
Fuchs, T. (2018). *Ecology of the brain: The phenomenology and biology of the embodied mind.* Oxford University Press.
Gallagher, S. (1997). Mutual enlightenment: Recent phenomenology in cognitive science. *Journal of Consciousness Studies, 4*(3), 195–214.
Gallagher, S. (2005). *How the body shapes the mind.* Oxford University Press.
Gallagher, S. (2012). What is phenomenology? In S. Gallagher (Ed.), *Phenomenology* (pp. 7–18). Palgrave Macmillan UK.
Gibson, J. J. (2015). *The ecological approach to visual perception.* Psychology Press.
Heidegger, M. (1962). *Being and time.* (J. Macquarrie & E. Robinson, Trans.). Blackwell Publishers.
Heras-Escribano, M., & de Jesus, P. (2018). Biosemiotics, the extended synthesis, and ecological information: Making sense of the organism-environment relation at the cognitive level. *Biosemiotics, 11*(2), 245–262.
Heras-Escribano, M., Noble, J., & de Pinedo, M. (2015). Enactivism, action and normativity: A Wittgensteinian analysis. *Adaptive Behavior, 23*(1), 20–33.

Husserl, E. (1900). *Logical investigations.* (J. N. Findlay, Trans., D. Moran, Ed.) (Vol. II). Taylor & Francis.

Husserl, E. (1970). *The crisis of European sciences and transcendental phenomenology: An introduction to phenomenological philosophy.* (D. Carr, Trans.). Northwestern University Press.

Husserl, E. (1982). *Ideas pertaining to a pure phenomenology and to a phenomenological philosophy I.* (F. Kersten, Trans.) (Vol. II). Kluwer Academic Publishers.

Husserl, E. (1989). *Ideas pertaining to a pure phenomenology and to a phenomenological philosophy II.* (R. Rojcewicz & A. Schuwer, Trans.) (Vol. III). Kluwer Academic Publishers.

Husserl, E. (1991). *On the phenomenology of the consciousness of internal time (1893–1917).* (J. B. Brough, Trans. Kluwer Academic Publishers.

Husserl, E. (2001). *Analyses concerning passive and active synthesis: Lectures on transcendental logic.* (A. J. Steinbock, Trans.). Kluwer Academic Publishers.

Legrand, D. (2006). The bodily self: The sensori-motor roots of pre-reflective self-consciousness. *Phenomenology and the Cognitive Sciences, 5*(1), 89–118.

Lutz, A., & Thompson, E. (2003). Neurophenomenology integrating subjective experience and brain dynamics in the neuroscience of consciousness. *Journal of Consciousness Studies, 10*(9–10), 31–52.

Merleau-Ponty, M. (1942). *The structure of behaviour.* (A. L. Fisher, Trans.). Duquesne University Press.

Merleau-Ponty, M. (1964). *Le visible et l'invisible: suivi de notes de travail. par Maurice Merleau-Pont. Texte établi par Claude Lefort.* Gallimard.

Merleau-Ponty, M. (2012). *Phenomenology of perception.* (D. Landes, Trans.). Routledge.

Morris, D. (2004). *The sense of space.* State University of New York Press.

Morris, D. (2018). *Merleau-Ponty's developmental ontology.* Northwestern University Press.

Noë, A. (2004). *Action in perception.* MIT Press.

O'Regan, J. K., & Noë, A. (2001). A sensorimotor account of vision and visual consciousness. *Behavioral and Brain Sciences, 24*(5), 939–973.

Rolla, G., & Huffermann, J. (2021). Converging enactivisms: Radical enactivism meets linguistic bodies. *Adaptive Behavior, 30*, 345.

Thompson, E. (2004). Life and mind: From autopoiesis to neurophenomenology. A tribute to Francisco Varela. *Phenomenology and the Cognitive Sciences, 3*(4), 381–398. https://doi.org/10.1023/B:PHEN.0000048936.73339.dd

Thompson, E. (2005). Sensorimotor subjectivity and the enactive approach to experience. *Phenomenology and the Cognitive Sciences, 4*(4), 407–427.

Thompson, E. (2007). *Mind in life: Biology, phenomenology, and the sciences of mind.* Harvard University Press.

Thompson, E. (2016). Introduction to the revised edition. In F. Varela, E. Rosch, & E. Thompson (Eds.), *The embodied mind: Cognitive science and human experience* (2nd ed.). MIT Press.

Thompson, E., & Stapleton, M. (2009). Making sense of sense-making: Reflections on enactive and extended mind theories. *Topoi, 28*(1), 23–30.

Toro, J., Kiverstein, J., & Rietveld, E. (2020). The ecological-enactive model of disability: Why disability does not entail pathological embodiment. *Frontiers in Psychology, 11*(1162). https://doi.org/10.3389/fpsyg.2020.01162

Varela, F. J. (1991). Organism, cognitive science and the emergence of selfless selves. *Revue européenne des sciences sociales, 29*(89), 173–198.

Varela, F. J. (1996). Neurophenomenology: A methodological remedy for the hard problem. *Journal of Consciousness Studies, 3*(4), 330–349.

Varela, F. J. (1999). *Ethical know-how: Action, wisdom, and cognition.* Stanford University Press.

Varela, F. J., Thompson, E., & Rosch, E. (2016). *The embodied mind: Cognitive science and human experience* (2nd ed.). MIT Press.

Von Uexküll, J. (1934). *A foray into the worlds of animals and humans: With a theory of meaning.* University of Minnesota Press.

Wheeler, M. (2010). Minds, things and materiality. In L. Malafouris & C. Renfrew (Eds.), *The cognitive life of things: Recasting the boundaries of the mind* (pp. 29–38). McDonald Institute for Archaeological Research.

Zahavi, D. (2003). *Husserl's phenomenology.* Stanford University Press.

Zimmer, C. (2008). *Microcosm: E. coli and the new science of life.* Random House, Inc.

CHAPTER 5

The Ecological Dimension of Sense-Making: The Environment as an Active Ecological Field

In typical accounts of sense-making, the constitution of vital and sensorimotor norms has two fundamental sources: the living body and its physical surroundings. The autonomous organization of the body determines the norm that modulates the interactions with the environment. The materiality of the environment facilitates or constricts the actions of the body and the possible norms to be enacted. For this reason, if we compare the relationship of the agent and the environment to the roles a couple plays when dancing, enactivists say the agent would be the leader and the environment the follower. However, this description is inaccurate if, as I argued in Chap. 4, sense-making consists of processes of norm development.

The thesis of norm development tells us that sense-making does not imply the emergence of an *Umwelt* from the bare materiality of the environment, and the autonomous organization of the body. Instead, sense-making consists of reconfiguring the pre-given normative field that has been enacted in the past of the agent–environment system. In this view of sense-making, there is already a normative bond between the body and the environment. When a new form of sense-making occurs, the environment constrains the actions and *enactions* of living agents, both normatively, and materially. Normative constraints shape the system's development, and in this sense, the environment plays a more active role than enactivists affirm.

M. A. Sepúlveda-Pedro, *Enactive Cognition in Place*, New Directions in Philosophy and Cognitive Science,
https://doi.org/10.1007/978-3-031-20282-7_5

This chapter will argue that in sense-making, an agent interacts with the environment as an active ecological field of action, and not simply as a bare material landscape. The argument proceeds as follows. First, I unveil the extreme passivity the enactive approach attributes to the environment when its supporters describe it as a material landscape (Sect. 5.1). From there, I develop an account of sense-making as something akin to creative improvisation that uses multiple environmental structures and dynamic interactional forces that ultimately surpass the solitary relationship between an autonomous system and its surroundings (Sect. 5.2). Having considered these ecological aspects of sense-making, I propose an understanding of the enactive environment that is similar to descriptions of contemporary ecological approaches to cognition that derive from Gibson's theory of affordances (Sect. 5.3). There is, however, a problem of incompatibility between the enactive approach and these Gibsonian models of cognition that we need to address. Ecological psychology posits that meaning is intrinsic to the environment and disregards the crucial role that the concrete development of an agent–environment system plays in constituting and transforming an ecological field. Consequently, I offer an alternative account of the ecological field based on Merleau-Ponty's concept of spatial levels (Sect. 5.4.). This concept allows us to understand the continuous development of norms in the temporal unfolding of the body and the world, and thereby gives us both the enactive and the ecological view we require to understand sense-making as a process of norm development.

5.1 The Broom Dancing Metaphor

The classical description of sense-making describes the addition of meaning and value to raw physical aspects of the environment by living agents, thanks to their vital and sensorimotor activities. Chapter 3 argued extensively against this description that I called the thesis of significance as a surplus. There, I suggested that we should understand sense-making as a process of norm development. Namely, sense-making is a process of *re*shaping or *re*configuring an *already existent* normative entanglement of the body and the world.

The enactive approach increasingly adopts descriptions and explanations of sense-making that can fairly qualify as norm development. Nonetheless, it is still common to misunderstand the role the environment plays in the constitution of sense-making norms. This section argues the *broom dancing metaphor* is the best illustration of the current conception of the environment from the perspective of the enactive approach. This metaphor implies that the environment of a living organism is nothing but a set of material constraints that both enable and limit the bodily actions of agents, as when a dancer uses a broom in her dancing performance. I will argue later this metaphor is wrong-headed because it fails to capture how sense-making happens in the agent's fields of action.

This section first depicts the view of the enactive approach about the relationship between an agent and the environment in processes of life and cognition. Supporters of the enactive approach accept to illustrate this relationship with the metaphor of a couple dancing together. Still, they claim the agent plays the role of the leader and the environment the role of the follower (Sect. 5.1.1). In what follows, I show why this metaphor does not work for understanding the relationship between the agent and the environment (Sect. 5.1.2). Finally, I show that Thompson and Stapleton's descriptions resemble a scenario where a solitary performer handles a broom while dancing alone (Sect. 5.1.3).

5.1.1 The Couple Dancing Metaphor

For the enactive approach, processes of life and cognition entail the systemic unity of a living agent with its environment. In the interactional dynamics of this coupled system, however, the agent and the environment play different roles because they maintain an asymmetrical relation (Thompson & Stapleton, 2009). This asymmetry is exemplified by the role they play in the constitution of norms of sense-making.

The purposiveness of norms, and the capacity to modulate the agent–environment coupling in favor of these norms, have their genesis in the autonomous organization of the living body (Di Paolo & Thompson, 2014). Vital norms arise from the biological needs of living organisms because these organisms *must* maintain the viability of their autonomous organization. Additionally, sensorimotor norms are also enacted thanks to the capacity of the living body to self-organize and perform actions that sustain a repertoire of bodily habits (Di Paolo et al., 2017). As we have seen (Sect. 4.3), the processes of adaptivity that these interactional norms involve and the eventual emergence of new norms from older ones are possible because of the self-organizational skills of the living body. In its minimal form, agency precisely entails the capacity of living systems to modulate their behavior and change environmental circumstances according to their own purposes (Sect. 3.4.4).

The active role of agents in the constitution and development of norms contrasts nonetheless with the "passive role" that supporters of the enactive approach attribute to the environment, either as *Umgebung* (physical surroundings) or as *Umwelt* (meaningful world) (see Sect. 4.1). As *Umgebung*, the environment is a set of physical processes that can constrain, shape, and alter the causal constitution and development of the living body and its repertoire of sensorimotor skills. The *Umgebung* plays no active role in the constitution of norms and merely facilitates and limits the enactment of sense-making norms. As *Umwelt*, things are not much better. Described as the specific region of the physical world that matters for the activity of living organisms, the *Umwelt* appears as an *already constituted* world of significance. In other words, the *Umwelt* results from enacting norms but has no *constituting* normative force.

A simple metaphor can help us see the conception of the environment that appears in many of the enactivist literature. Clark (2008), in his own vision of embodied cognition, compares the relation of the agent and the environment in cognition to a couple dancing together, interacting and mutually constraining the movements of each other. The agent and the environment are in this picture like two equal partners.

If we take this metaphor in the enactivist picture, Thompson and Stapleton (2009) ask us to conceive the agent as the one leading the dance precisely because the agent modulates the coupling to fulfill the norms guiding its behavior. Although, I agree with Thompson and Stapleton on this clarification. I think there are still some problems that we can illustrate from this way of picturing the agent–environment relation in sense-making processes. I examine these problems in the next section.

5.1.2 Dancing with Others

There are two main problems with the couple dancing metaphor for illustrating the relationship between an agent and the environment within the framework of the enactive approach. First, the enactive approach describes the environment as a much more passive participant than any real dancing partner in a dance performance. Second, we cannot compare the agent–environment interactions to those between two living agents. These latter interactions involve emergent levels of normativity that simply cannot be present in or emerge from a relation between a living and a nonliving being. Let me explain these two issues more broadly.

Dancing in a couple requires that two agents must actively move their bodies. One of the agents might be on the lead, guiding the interactional coordination of the two bodies, but the other partner does not merely constrain the leader's movements; the follower is not a mere burden nor a simple limitation for the bodily actions of the leader. The follower actively changes and alters the dance flow, and her contribution to the performance is no less significant than that of the leader (van Alphen, 2014). A dance performance is, in this case, co-constituted by the interactive *normative* forces of the two agents.

The enactive approach describes the normative behavior produced by the interaction of two agents as *participatory sense-making* (De Jaegher & Di Paolo, 2007). A new layer of collective normativity likely emerges when two or more agents interact. This emergent normativity involves the proper coordination of the participants' bodies. Joint coordination requires that agents adjust their individual norms, which causes a primordial tension between two levels of normativity: the individual and the collective. This tension is permanent, no matter how adequate the level of coordination may be, because the participants remain autonomous. When the bodily coordination of agents succeeds, a new type of norm, a basic social norm, is enacted.

Like any other norm that emerges from a sense-making process, these norms are yet not crystalized. Instead, they involve a continuous process of actualization that may sometimes lead to enacting new norms. This phenomenon happens in dancing, where the bodies of the two participants constantly adjust their behavior to accomplish the proper collective coordination. Significantly, these processes of adjustment and adaptivity happen so long as an interactional coordination exists. The "right" coordination between two or more agents involves a *co-regulation* of individual and collective sensorimotor norms that allows agents to accomplish a good performance (Di Paolo et al., 2018).

The primordial tension, the emergence of collective norms, and the coregulation processes are features of interactions between living agents that are not present in interactions between a single individual and nonliving beings (De Jaegher, 2018). The problem with the dancing couple metaphor, which Thompson and Stapleton mistakenly follow from Clark's suggestion, is that the role of the environment as *Umgebung* is too passive. Conceived as a mere set of physical constraints, the world is a burden for the activity of agents, which are, in this story, the only active constitutors of their normative realm. For this reason, the picture one gets from this metaphor is more that of a single person dancing with a broom, or more accurately, swinging a broom around as they make dancelike motions. They are not strictly speaking engaged in a dance with the broom since the broom is not dancing at all.

5.1.3 Dancing Alone

Dancing with a broom illustrates better the existing relationship between an agent and a nonliving thing in bodily practices from the current standpoint of the enactive approach. It involves one person manipulating an object while dancing alone. A thing (the broom) is nothing but a physical constraint the agent (the dancer) needs to deal with to perform her action (dancing).

In a performance consisting of a dancer with a broom, the physical constitution of the broom can enable or constrain specific bodily movements of the dancer, but the adaptive behavior of the dancer handles these constraints. The capacity of the dancer to accomplish her performance correctly can be put in terms of the dynamic models of the enactive approach based on Piaget's theory of equilibration.

According to the dynamical account of Piaget's theory of equilibration (Sect. 4.2.2), specific aspects of the environment are modeled as variables that either reinforce or perturb the dynamical behavior of sensorimotor schemes. Hence the environment is a set of physical constraints that cooperates to sustain the autonomous organization of a sensorimotor habit and motivates the reconfiguration of this habit through processes of adaptivity or contributing to the enactment of a new habit that either coexists with the older habit or replaces it (Di Paolo et al., 2014).

Likewise, dancing with a broom can challenge the already existing repertoire of bodily movements of the dancer. The broom needs to be assimilated by the dancer so that her actions acquire a new normativity when she accommodates her movements to manipulate the broom. In Piaget's language, the broom is incorporated into the sensorimotor skills of the agent.

There is, however, a more interesting phenomenon that may take place in the interaction between agents and nonliving objects: tool-incorporations. This phenomenon has been described by Merleau-Ponty (2012) and Bateson (1987), recognized by the enactive approach (Di

Paolo, 2009) and is supported by empirical evidence (Bach-y-Rita & Kercel, 2003). These incorporations do not simply mean that an agent can accurately handle a worldly object, as with Piaget's theory. In bodily incorporations, a tool such as the cane of a blind person or the tennis player's racquet can be incorporated into the body's agency to extend its repertoire of bodily movements. Learning to dance with a broom entails the possibility that dancers incorporate the broom like the blind person incorporates a cane. In short: The broom can, in principle, flow all along with the bodily movement of a dancer, as if it were an extension of her body.

Here, the difference between both metaphors shines through, for dancing with another agent does not entail the possibility of incorporations, at least not in the way a tool or an object is incorporated.[1] One dancer cannot be incorporated by another simply because the autonomy of both agents causes an insurmountable tension that impedes the level of submission of one body's movements to another's movements, as can happen with a tool. Only if one of the bodies loses its autonomy is it possible, in theory, to become incorporated as a tool by the other body.

Although the interactions between a living agent and nonliving objects are not the same as those between two agents, this does not mean that the environment needs to be as passive as Thompson and Stapleton have it. They rightly say the body–world relation is not symmetrical. If we accept the theory of biological autonomy, we must assume that living organisms are agents that unavoidably constitute their purposes, and they can actively transform their interactions with the environment. Nevertheless, the environment has a more "active" role than enactivist acknowledge if we realize the environment an agent interacts with, in processes of life and cognition, is an *ecological and normative field of action.*

Looking more carefully at processes of life and cognition, we can realize that the environment an agent encounters in its sense-making activities does not only consist of a set of material constraints that facilitates or constricts the actions and purposes of agents. The enactive environment is not a bare material landscape. By contrast, for the agents, the environment is a rich domain of forces that shape and push their development, inviting and inhibiting the enactment of new sense-making norms. Multiple normative and ecological forces are at play in an agent's field of action. In the next section, I mention three aspects of the environment that reveal how enacting a new sense-making norm entails more than the autonomy and purposiveness of the agent and the material constraints of things.

5.2 The Environment as an Ecological Field of Action

The broom dancing metaphor does not adequately capture the agent–environment relation in life and cognition because it is based upon the same mistaken assumptions of the thesis of significance as a surplus (see Sect. 4.1). It looks at the environment as a purely physical and neutral value domain that we can conceive, at some point, as something separated or independent from the historical development and evolution of the body–world entanglement. Since partisans of the enactive approach are committed to a view of cognition and science that implies the fundamental circularity (Sects. 3.1.2 & 4.1.2), they must be aware that thinking of the environment as a purely physical world is nothing but an abstraction. Indeed, this abstraction could be a valuable and necessary presupposition of our natural attitude for doing science (see Sect. 4.1.3). The problem is that such presupposition obscures our understanding of sense-making.

In Chap. 4, I argued that sense-making is a process of norm development. Such a process implies that the body and the world are always entangled, which means there are always previous normative elements at work in enacting new norms, that is, in sense-making. If so, the environment never appears as a raw or purely physical domain. It rather appears as a setting already constituted by some normative aspects that prefigure and constrains the sense-making of agents.

The notion of *the environment as an ecological field of action* that I suggest can, however, better account for the environment's role within the agent–environment system's normative domain. This notion describes the environment as an active ecological field of forces that constrain, ***normatively and not only materially***, the development of the agent–environment system. As such, it also motivates the self-transformation of the normative domain of this system because of the ever-present disparity between the habitual self-organization of the body and the ongoing changes in the normative realm of the world. This new conception of the environment as an active ecological field of action thesis fits better with the thesis of norm development and helps us understand the self-transformation of the body–world entanglement as an enactive and ecological process.

This section illustrates three fundamental aspects of the environment relevant to sense-making that enactivists do not capture with their dualistic account of the environment (i.e., the *Umgebung–Umwelt* duality). First, the hypothetical example of a human baby learning to bring the attention of her parents exemplifies the enactment of a new norm of interaction from unexpected environmental circumstances (Sect. 5.2.1). These circumstances are constituted by the predisposed responses of other agents to the fortuitous occurrence of an unintended sensorimotor action and not by the purposive intentions of the baby. This example exhibits the importance of chance and creativity for the enactment of new norms of sense-making. Therefore, I argue that sense-making is an act of creativity akin to the practice of jazz improvisation (Sect. 5.2.2). There, successful musicians bring forth new ways of musical sense-making actualizing and reconfiguring their habitual interactions in light of the current conditions of the environment. Jazz improvisation is a paradigmatic example of norm development, but, as I will argue in the same section, actualizing and reconfiguring are indeed the hallmarks of sense-making. I use the same example to illustrate the richness of the ecological and historical dimensions of musical sense-making (Sect. 5.2.3).

5.2.1 *Causal Laws and Normative Constraints*

The thesis of significance as a surplus describes the addition of meaning to a raw physical environment that doubles the references we make of the surroundings of living agents: as *Umwelt* (or lived environment) and as *Umgebung* (or the subject-independent world described by sciences like physics and chemistry).

I argued in the last chapter that such a description does not entail any kind of traditional dualism that radically separates the subjective sphere of living agents from the theorized subject-independent world. Instead, it involves a phenomenological dualism that shows the environment from two different experiential-embodied perspectives of human beings. The error of the thesis of significance as a surplus is to hypothesize the origins of the subjective mode of experience and its counterpart, the lived world, in the contents of the intersubjective-enculturated mode of experience, understood as the objective world described by science. The same error undermines the broom dancing thesis.

In the broom dancing thesis, the active role of the agent involves the superimposition of a normative layer onto a physical domain that is completely passive at the normative level. The only "active" part of the environment is found at the causal or pre-normative levels of the agent–environment system. On this level, the environment enables and constrains the development of the living organism. However, once it is invested by meaning, the environment becomes nothing but a lived world with no influence or determination over the enactment of new norms of sense-making.

Consider the example of the infant that incorporates the sensorimotor habit of feeding on a bottle of milk instead of, or additionally to, breast-feeding (Sect. 4.2). The bottle reshapes the agent–environment system's normative domain, exposing the baby's motor actions to a new set of physical constraints that differ from those of the mother's breast. The dynamical model offered by the enactive approach of this incorporation illustrates how the environment causally reshapes the sensorimotor dynamics of the baby and motivates the enactment of a new norm, a new sensorimotor contingency.

The environment as a causal determination (as a physical constraint) never disappears; it alters the agent's sensorimotor dynamics. As long as the baby consumes milk from the bottle, for instance, the weight of the bottle will change. As a result, the pressure and tension of the muscles in the baby's hands and arms will progressively change to accommodate her body and satisfy the sensorimotor norm that sustains her feeding. Holding the bottle in new environments, on the floor instead of on the bed, and taking different postures when feeding on the bottle, for example, seated instead of lying down, will also alter the original sensorimotor norm. The environment never stops challenging an agent and remains *causally* active. However, as Thompson and Stapleton affirm, the agent keeps the leadership of the dance because it is the baby who adjusts the sensorimotor loops to fulfill the norm of her behavior.

Understood as a meaningful aspect of the infant's *Umwelt*, the bottle will never stop motivating new sensorimotor norms. The bottle of milk first affords grasp-ability, suck-ability, and the opportunity for feeding. Later, for some babies, the same bottle can begin to afford new behaviors and satisfy new sensorimotor and vital norms, even without the necessity of an initial purposiveness of the agents.

While holding the milk bottle, some babies can start hitting another object with the bottle, causing noises. They can also learn to throw the bottle away and thus bring the attention of their caregivers, finding a new practical use of their recently acquired sensorimotor repertoire. If such a thing happens, a new sensorimotor norm is enacted, and its enactment does not depend on the agent's initial purposes or the environment's raw physical constraints. The new norm is grounded on the already acquired sensorimotor repertoire (a normative framework) that finds its realization in a new and unexpected circumstance or situation. An unexpected normative attunement is achieved, and the enactment of a new skill is incarnated in the body as a new "I can" and in the environment as a new "possibility for action" (an affordance of the bottle), given the appropriate circumstances.

As we see in this example, the environment plays a more active role than the broom dancing thesis captures. As a motor signification or a unit of affordances, the bottle actively contributes to enacting a new norm. Its role, in other words, is not merely causal. Grasping the bottle never occurs without a set of physical constraints, but the new behavior of the baby does not simply arise from incorporating such constraints in a new fashion. Instead, the previous norm, the grasping of the bottle for feeding, works as a scaffolding that allows for the enactment of a new sensorimotor behavior, a new sense-making norm.

From a phenomenological perspective, Morris (2004), inspired by the development of motor skills in infants, offers a helpful metaphor for the self-transformation of the normative domain of the agent–environment system: origami. In origami, we make a series of folds on a piece of paper to create a determinate figure that can resemble an animal or a familiar object. An origami's correct performance consists in properly folding the paper to achieve a fold structure showing the figure we wish to represent. Correctly folding the sheet of paper entails a specific sequence in which each fold (save for the first) is performed under the constraints of previous folds, which produce new constraints for subsequent folds, etc. However, we cannot say that one fold *determines* the next; instead, each fold limits the number of possible folds that may follow. These folds are not a mere limitation or negative result of the performance. Instead, they are open possibilities for making our wanted figure or improvising and creating a new figure (Sawyer, 1992). This example leads us to another aspect of the environment, which is more than physical constraints and actively motivates the enactment of new sense-making norms: the ecological field of forces.

When an infant realizes she can make noise and get attention from others by hitting a bottle of milk against another object or throwing the bottle away, she finds her actions fit into a new field of forces that solicits her action in a new way. This field is not subject-independent since it is in her interest to get her parent's attention. She does not constitute the field either; the field instead emerges from the new reconfiguration of the whole agent–environment system, which finds a new orientation or a new sense. The disparity between the habitual past (the bottle-feeding norm) and the newly enacted present (the bringing attention with the bottle) is not the result of incorporating something senseless into the agent–environment system's normative domain. Instead, it is a whole reconfiguration of the agent–environment system in a creative process that involves the self-transformation of the body and the world, all at once, where these self-transformations are also importantly informed by their past.

Creativity and improvisation are indeed some of the most essential aspects of sense-making. The following subsection will argue that creative improvisation involves a dialectical movement of the body and the world akin to the situation I have described above.

5.2.2 *Sense-Making as Creative Improvisation*

Improvisation and creativity are fundamental aspects of life and cognition. Contrary to our commonsense view of these phenomena, they do not spontaneously bring forth something totally new but instead consist of taking up something that is already done in a new way (Montuori, 2003). Jazz musicians create improvisations by adapting a traditional melody (a standard), a habitual repertoire of melodic phrases, and harmony shifts so that new combinations bring forth unique performances. The enacted performance of improvisation brings forth the habitual repertoire into the field of forces constituted by the social and musical environment of the performance (Walton et al., 2015).

Crucially, the improvisation does not originate from the intentions of the musician alone. Playing with a band in public causes a series of contingencies that alters any preestablished plan of the musician (Sawyer, 1992). Hence, improvising consists of working with these contingencies to bring forth the old repertoire anew, like creating a new figure by improvising folds from the already known sequences of folds in origami.

Therefore, the accomplishment of good jazz improvisation entails a new balance between the old and the new, between the habitual past and the unexpected contingencies of the present. The dynamics of agent–environment and past–present occur at multiple levels. At the causal level, the physical constraints of the environment cause a constant tension that agents must mediate. In jewelry making, for instance, goldsmiths need to constantly adjust and adapt previously acquired techniques, depending on the type of "responses" the material sends back to their bodily actions (Baber et al., 2019). This dynamic process of an endless readaptation of their actions designates the first set of tensions that force agents to reshape their acquired bodily habits to suit environmental conditions constantly. Consequently, skillful jewelry making is more a capacity to deal with the unexpected condition of matter than knowing in advance how to give form to passive matter.

Di Paolo et al. (2018, p. 141) have already recognized a primordial tension between the agent and the environment due to a permanent "messiness" of the materiality of bodies. It must be noticed, however, that the material challenges of things are not intrinsic to their material composition. The messiness results from the failed or partially failed agents' expectations. These expectations are rooted in sense-making norms. Materials resist against the hand that aims to manipulate them when a norm of action guides the hand. Without habitual ways to manipulate materials, there would be no unfulfilled expectations. Understood this way, the world forces the agent–environment system to actualize its normative domain indefinitely, resulting in the unpredictable and unintended consequences of body–world interactions.

5.2.3 Environmental Structures

The tensions and forces of the environment challenging the agent are not only originated in its materiality and normative expectations. The interactions with other agents also produce a rich field of tensions that moves forward the agent–environment relationship, like in the example of the baby finding a new whole context for her sensorimotor skills. This is like the creation of new personal and collective styles of playing music.

In jam sessions, improvisation entails a constant self-transformation of interactional norms, but this practice's dynamic "self-movement" goes further. In these sessions, musicians constantly and collectively create new structures of sense from previously given structures of sense (musical structures). Some new musical structures are felt as successful and become part of the habitual repertoire of one or more practitioners. The accumulation of this new musical repertoire can eventually transform each musician's current personal norms and the norms of the collectivity, either of a small group or a whole cultural society (Sepúlveda-Pedro, 2020).

For musical creativity, the adaptive behavior of the musician is undoubtedly crucial, but the creative process involves the participation of multiple forces. The most obvious is the contribution of the other musicians. This is why jazz improvisation involves participatory sense-making (Schiavio & De Jaegher, 2017), similar to the practice of a couple dancing. However, in the case of music, there are not only bodies interacting directly in the performance, but also a whole environmental structure involved and co-constituted with other agents. The environmental structure creates "folds" and scaffoldings that enable and constrain the development of the improvisation and, eventually, of a whole new musical style.

These environmental structures can be musical patterns, co-constituted and co-enacted with other musicians that no longer play a passive role in music production. These structures do not merely constrain the musician; they solicit the musician's action in a certain way (cf. Walton et al., 2015), leading the agent to find a new creation, a new field, or even a new musical *Umwelt*. Therefore, the musician navigates in the normative field of forces co-constituted, co-enacted by multiple agents, and embodied in the network of meaningful environmental structures that enable and constrain her possible actions.

We should notice that the examples of jazz improvisation and jewelry making involve a sociocultural dimension. The social dimension of cognition is a phenomenon that I will explore more carefully in Chap. 6. Here, I wish to illustrate the presence of a normative field always present in the body–world entanglement, one that shows the real presence of the environment in cognition. For instance, in the example of the bottle of milk, the baby applies creativity and improvisation to enact new norms and finds new configurations for old norms in the new field of forces. The new norm does not come out directly of the intrinsic teleology of the baby, not even from her sensorimotor habits. Instead, it arises from reenacting old habits in a new ecological field.

The ecological field is the normative dimension of the environment, which is not just a subject-relative field. The *environment* is also the horizonal structure of the world, and it is constituted by a complex interrelation of things, agents, and the body itself, something I will name enactive place in Chap. 6. As such, this active environment does not correspond to the abstract domains described by physics and chemistry; instead, the environment is a place of dwelling (Malpas, 2004), or more simply, an ecological niche constituted by a web of interrelations between multiple agents, things, mediums, surfaces, and so on. As we will see in the following, this is precisely where the process of sense-making as norm development takes place.

5.3 The Ecological Dimension of Sense-Making

From a phenomenological perspective, the first dimension of our perceptual world is ***depth*** because the world is lived primarily not as an abstract set of physical determinations but as a place of dwelling (Casey, 1996; Morris, 2004). Dwelling is, therefore, the condition of an agent situated in a setting of already preestablished normative relations that its body maintains with other living and nonliving aspects of its surroundings. This scenario of the environment as an ecological field of action assumes that the agent is situated in a relational field that I will call *the ecological dimension of sense-making*. In this ecological dimension, any norm of sense-making is enacted, which is why we need to specify the fundamental characteristics of this dimension.

The ecological dimension of cognition has been recognized not only by phenomenologists but also by Gibson's ecological approach to visual perception. This section will first review Gibson's approach (Sect. 5.3.1). The enactive approach has been reticent to endorse some essential ideas of Gibson, like the theory of affordances and the theory of ecological information, because he claims that these elements of the ecological field are subject-independent (Sect. 5.3.2). This problem is surmountable, nonetheless, if we acknowledge that the ecological field is the presence of a normative past enacted in the history of the agent–environment system and not a set constituted by causal processes that are external to the concrete existence of this system. From this standpoint, Chemero's dynamical interpretation of the theory of affordances helps us to bring ecological and enactive approaches closer (Sect. 5.3.3). However, Merleau-Ponty's phenomenology of spatial levels provides us with the best conceptual framework to conceive the ecological dimension of sense-making from a full-fledged enactive standpoint (Sect. 5.4).

5.3.1 Gibson's Theory of Direct Visual Perception

Gibson's ecological approach has significantly influenced radical embodied cognition because his theory offers a solid scientific framework for explaining perception without appealing to internal representations. For Gibson, our active exploration of the environment provides us with all the information we need to account for perception. Importantly, this information is not theoretical but practical because it concerns what the environment affords (for good or for ill) as possibilities for action on the part of the perceiver (Gibson, 2015, p. 119). These possibilities for action are called *affordances*.

Although affordances are not, as such, physical properties of things, they are nonetheless objective features for Gibson in the sense that they exist regardless of the factual presence of any perceiver. However, Gibson (2015, p. 121) also claims that affordances are neither objective nor subjective because they are beyond such a simple dichotomy. He based this claim on the idea that affordances pertain to the relational domain of the animal–environment system. My coffee cup, for instance, has the affordance of graspability. This affordance does not exist in my head but in the cup itself, yet it is not a property of the cup as such but is also relative to the body that can grasp it. Consequently, the cup has the affordance of grasp-ability for a human or maybe some other primate, but it does not offer the same affordance for a dog or a snake.

For this reason, some ecological psychologists after Gibson claim that affordances are more appropriately understood as relational properties of the animal–environment system (Heft, 2015). We can "objectively" study these relational properties from a phenomenological stance because they are, in principle, as observable and measurable as natural physical phenomena. The real challenge, for Gibson, is to explain how organisms can perceive affordances without internally representing any feature of the outside world. Gibson's hypothesis, in this regard, is that some kind of information in the environment facilitates the perceptual recognition of affordances. This is what Gibson (2015) calls *ecological information*.

Ecological information, like affordances, does not refer to intrinsic physical properties of the environment, such as sensory data or physical stimuli. For Gibson, sensorial data do not convey any information about the environment. Although stimuli may cause sensorial impressions in our perceptual systems, they can be structureless, which would tell us nothing (Gibson, 2015). Think about a well-illuminated room full of fog. The fog causes the reflectance of light in all directions, creating a uniform white color all around the perceiver that impedes her from seeing the structural conditions of the room. As Gibson claims, the uniformity of stimuli is equal to the total absence of stimuli (e.g., a dark room).

Ecological information implies a "structure" of matter and energy typical of habitats on Earth. This structure exhibits variant and invariant elements but is relative to animals' locomotion (Gibson, 1966). The ambient optic array is the most typical example of ecological information offered by Gibson. Usually, light does not impact our retinas directly from the source of light (e.g., the sun or a light bulb). Light is spread all over the places and landscapes we inhabit from the source, causing different effects of reflectance. The reflected light of a visual scenario that Gibson called ambient light converges to every possible standpoint an observer can occupy (or not). When the standpoint is occupied, the observer can already perceive differences in the different zones of the visual scene. If the observer changes her perspective, she can appreciate that the visual scene also changes. Later, when she returns to her original position, the first visual scene returns. We can observe some aspects of the environment remaining invariant in the transition from one scene to another, such as the corner angles of a table. The body itself can appear as one of these structural invariants, like our nose, that is always part of our visual field. The invariants in the ambient optic array's flux help us anchor our movements' orientation and give a sense (a direction) to the visual scenes we perceive. Gibson compares the role of these invariants to the presence of the Parthenon in Athens (Gibson, 2002). The Parthenon, visible from every city's point, helps us know our position while wandering the city.

We should notice that the variant– invariant structure of the environment or ecological information is not something our minds make up. We only need to disclose this environmental structure through our locomotion. This is a process that Gibson called *information pickup*. Ecological information is outside individuals' heads, and it is, for this reason, observable and measurable by science. Although Gibson could not prove that ecological information is lawful, his followers later found in differential equations (Warren, 1984) and models of dynamic systems theory (Turvey & Carello, 1995) empirical solid support for Gibson's hypotheses.

Gibson's claims have important convergences with the enactive approach (Chemero, 2009) and other forms of enactive cognitive science (Hutto & Myin, 2017). However, there are also significant differences. Section 5.3.2 surveys the convergences and discrepancies of these two forms of world-involving cognition and problematizes Gibson's conception of affordances as subject-independent features of the environment.

5.3.2 Enactive or Ecological Information?

The meanings perceived in the enactive *Umwelt* of living agents are easily comparable to the descriptions of affordances. Glucose affords nutrition for E. coli bacteria, while other chemicals like gold can afford intoxication to the same organisms (Villagrán et al., 2020). Likewise, in sensorimotor habits, objects in our surroundings afford possibilities for motor action. Stairs afford climb-ability, and the chair affords sit-ability.

However, for the thesis of significance as a surplus of the enactive approach, affordances imply adding value to things thanks to the vital and sensorimotor norms enacted by agents' bodies in their history of interactions with the environment. For Gibson, by contrast, affordances are located in the environment, and the factual presence of an agent is unimportant. Organisms only pick up ecological information available in the environment and get access to the affordances available in it. These two opposite views have thus far been the major difference between the Gibsonian school and the enactive approach (Di Paolo, 2016).

To be fair, Gibson's claims are problematic and confusing because he describes the visual field as a meaningful, subject-independent domain. This claim does not seem to fit with physicalist ontologies of science that presuppose the subject-independent world as a meaningless world. Gibson's aim, nevertheless, is obvious: He wants to make ecological psychology an "objective" science. Gibson's unusual characterization of the perceptual environment has also motivated endless discussions about the ontological status of affordances. Without discussing the ontological status of affordances but focusing on empirical evidence, we can undermine the subject-independent nature of affordances.

Mossio and Taraborelli (2008) compare the original theory of sensorimotor contingencies (O'Regan & Noë, 2001) and Gibson's ecological approach. These authors conclude that Gibson's approach misses the body's central role in enacting sensorimotor information. For Gibson, the variants and invariants of sensorimotor information are in the environment and can be disclosed by an agent's locomotion. Nonetheless, the specific bodily movements needed to reveal such information seems irrelevant to Gibson. By contrast, the theory of sensorimotor contingencies sees the self-organization of the body and brain as playing a crucial role in acquiring mastery of sensorimotor laws (Hurley & Noë, 2003).

Gibson emphasizes that we disclose the environmental structure of light thanks to locomotion. However, he does not acknowledge that such disclosure, in vision, depends on more than spatial displacements of perceivers. Rather the active movement of the body discloses visual information to an observer. From a phenomenological perspective, Susan Bredlau (2006) gives the example of how a person that has always been transported by car, but who has never learned how to drive a car, cannot perceive accurately the spatial information needed to accomplish this activity efficiently. The information required for driving is accessible to drivers who have incorporated the correlations between the active motion of their bodies (handling the wheel, pedals, gear, etc.) and the correlated flux of sensory information. Notably, there are phenomenological arguments and empirical evidence on this subject.

In an experiment with cats, two different groups of newborn cats are deprived of visual information (Held & Hein, 1963). Eventually, they are allowed to move in an illuminated environment but harnessed to a carousel. The first group of cats can move their legs along with the movement of the carrousel, while the second group is merely transported by the carrousel. After a period, both groups can move around freely in an illuminated environment. The cats of the first group can move efficiently in this environment, while the cats of the second group are clumsy and unable to walk correctly. These results suggest that perception and its development depend on the active sensorimotor coordination of agents and not solely on disclosing information via locomotion. That is, the active self-movement of the body is necessary to acquire the skills required to interact appropriately with the environment (Di Paolo et al., 2017).

Hence, ecological information is closer to what the enactive approach defines as a sensorimotor environment (an abstract space of sensorimotor laws that are independent of the active engagement of any agent) than it is to sensorimotor habits (the sensorimotor correlations that involve the emergence of habits as autonomous systems in precarious conditions). If so, agency and subjectivity are not relevant for Gibson's notion of ecological information, just as it happens in projects of weak enactivism.

Chapter 3 argued that the weak enactivism neglect of subjectivity and agency as constitutive aspects of cognition is problematic (Sects. 3.3 & 3.4). From this standpoint, a criterion for distinguishing truly cognitive agents and mere responsive mechanisms like those of embodied artificial intelligence was missing. It is true that we can deduce ecological laws and that mechanical systems can exploit these laws. However, there is no meaning or significance at play for the artificial mechanisms in these types of sensorimotor interactions. This lack of meaning is not a problem for the operation of an artificial mechanism because engineers foresee the sort of ecological information relevant to the machines. Therefore, the design of the device embodies the engineers' plan.

However, according to the theory of biological autonomy found in the enactive approach, the relevant ecological information for a living agent is the one that is vital for its autonomous organization. This means that the autonomous organization of a living agent is not entirely determined before its concrete interactions with the environment. Specifying what is relevant for the agent cannot be given a priori. In the interaction itself, some aspects of the organism's surroundings become relevant, correlative to its self-organization processes that are not predetermined. The enactment of a domain of significance or a normative domain corresponds to what we call sense-making, which distinguishes living from nonliving beings and cognitive from noncognitive beings.

Ecological laws are, therefore, scientific abstractions. While helpful, these laws are not part of the phenomenon of perception itself. Sensorimotor correlations are enacted rather than picked up.[2] They are part of the self-movement and self-transformation of the body–world entanglement rather than intrinsic characteristics of a subject-independent world. I will later argue that more contemporary accounts of affordances (Sect. 5.4.1) and Merleau-Ponty's notion of spatial levels (Sect. 5.4.3) describe this phenomenon. In what follows, I am looking at the problems arising from the claim that affordances are subject-independent features of the environment.

5.3.3 *Are Affordances Normative?*

For most ecological psychologists, affordances, like the structures of ecological information, are subject-independent. Affordances are relational and implicate the organism–environment system. However, these affordances are arguably constituted by causal processes that do not involve the active engagement of agents. For Turvey (1992), for instance, affordances are like dispositional physical properties. Organisms can react lawfully to the presence of specific characteristics of the environment, like how two chemical compounds react when put together in the appropriate context.

Similarly, Reed (1996) claims that affordances are relations that originate in processes of natural selection. Organisms respond to the affordances perceived in the environment not necessarily because they have developed a bodily affective sensitivity to the environment in their developmental unfolding in time. Instead, for this view, organisms are designed by nature to respond to affordances. Although affordances, in this case, imply the intimate ecological relationship between organisms and the environment, causal processes external to the agent–environment system development determine this relation. The enactive approach rejects this idea (see Sect. 3.4).

Similarly, Heras-Escribano and de Pinedo (2016) have argued that affordances are non-normative because normativity necessarily involves criteria of correctness that exist, as far as we know, only in the sociocultural practices of human beings. They base this claim on a Wittgensteinian account of normativity. Wittgenstein (2009) affirmed that sociocultural practices entail norms that have not only conditions of satisfaction but also conditions of correctness. A condition of satisfaction is when *E. coli* bacteria find the glucose they need for their metabolic requirements. As we saw in Chap. 3, *E. coli* can adapt its organization to metabolize lactose when the bacteria environment is low in glucose. In this case, either glucose or lactose can fulfill the vital norm of *E. coli*.

A correctness condition implies that we need more specific criteria to satisfy the norm. This happens, for instance, when a human being determines the quality of her food as good or bad. This judgment can be based not only on the food's nutrients but also on other more specific criteria such as a particular taste, the ingredients' origin, or the dish's aesthetic appearance. The existence of conditions of correctness is a common feature of human practices. They imply a new layer of constraints for human actions indeed. We must notice that individuals do not choose these criteria. Instead, they result from social conventions that individuals follow even to rule their solitary actions (Rietveld, 2008).

The main argument of Heras-Escribano and de Pinedo in support of a non-normative character of affordances is that affordances, as part of nature, are not as contingent as social norms are. For this reason, affordances depend on relations determined by causal processes of nature, social norms instead on conventions (Heras-Escribano, 2019).

This sort of argument has pushed many ecological approaches to distinguish between natural and social or cultural affordances (Ramstead et al., 2016; Rietveld & Kiverstein, 2014). Whereas bare biological processes can determine natural affordances, cultural affordances are more contingent and relative to the historical past of a given cultural group. While the distinction between natural and cultural affordances is undoubtedly useful, I disagree with the suggestion that natural affordances originate in bare causal processes independent of the normative domain of the body–world entanglement. The last section of this chapter will argue that affordances, whether natural or cultural, are part of the enacted normativities that constitute the agent–environment system. Before doing so, the Gibsonian distinction between the environment as a realm of raw physical processes and as a relational ecological domain needs to be clarified.

The distinction between the purely causal realm of physics and the ecological dimension of visual perception is already at play in Gibson, who distinguished between two kinds of science: physical and ecological optics. These two types of science look at two different aspects of the phenomenon. Physical optics analyzes the material composition of light and the causal effects of light on the visual system. Ecological optics, by contrast, studies the structure of light spread out in the different parts of the environment (medium, surfaces, and landscapes) that cause various effects of reflected light (e.g., transparency, color, and shape). The ecological realm, therefore, does not imply the analysis of light abstracted from the visual context. Instead, it analyzes the presence of light in the wild or how light is interrelated to the multiple objects and aspects of the environment (Gibson, 2015). Consequently, the relational domain of light, not the abstract realm of physics, gives perceivers the information necessary to recognize affordances.

However, the relational field is not restricted to the relation between light and the environment because the perceiver is also immersed in the system of relations. The perceiver needs to occupy a location in the ambient optic array and move through different arrays to disclose the invariants of the structure of light. Likewise, affordances are relations between organisms and the environment. The environment affords opportunities for action only to those organisms capable of exploiting such opportunities. Water affords breathing to animals with gills but not to animals with lungs. Flat ground affords walking for legged animals and crawling or slithering for animals without legs.

All this suggests we can find the origin of the relational field in the processes that constitute the bodies of organisms and in the intrinsic properties of the environment. As a result, an orthodox theory of evolution may explain the origin of the relational field, as Reed and others suggest. But things are more complex here. The last section of this chapter argues that the relational field described by ecological optics depends on developmental processes as much as on phylogenetical processes so that affordances are revealed as subject-relative via the phenomenon of norm development.

5.4 The Self-Transformation of the Body–World Entanglement

Ecological approaches generally assume philosophical and scientific views that supporters of the enactive approach find problematic as theoretical and empirical backgrounds for explaining cognition. Many followers of Gibson, for instance, see the ultimate causal bases of affordances in processes that are external to the concrete development of the agent–environment system. This view contradicts the enactive approach's philosophical thesis of embodied subjectivity (Sect. 3.1).

Furthermore, like radical enactivism (Sect. 3.3), ecological approaches attribute almost exclusively to processes of natural selection the causal origins of the ecological bond between an agent and its environment, neglecting the central role of developmental processes in the constitution of this bond. Recall that supporters of the enactive approach, in contrast, attribute the emergence of the ecological bond to the constitution of a living system as an autonomous system. Since the autonomy of a physical system is unsustainable without interacting with the environment adaptively and normatively, the link between the agent and the environment is determined by the autonomous organization of the agent's body and the material resources the environment provides it to sustain its process of self-maintenance. From this standpoint, evolution plays a significant role in shaping the body's structure and behavior of organisms' lineages. Still, it is in development that the inherited structures of the body and the environment become actualized according to the current condition of the agent–environment system (Sect. 3.4).

Putting aside discussions about these two different perspectives of life and evolution, my purpose here is to unveil the enactive nature of affordances. Such a perspective makes it possible to describe the enactive environment as an active ecological field of action, the primary purpose of this chapter.

According to the enactive view, affordances and ecological information cannot be structures independent of the concrete development of the agent–environment system. Instead, we should characterize them as normative and ecological structures enacted in the past of the agent–environment system, sedimented as a meaningful domain of interactions in the agent's niche. Hence, the enacted ecological structures would be relational and located in the environment, as ecological thinkers affirm. They would be indeed connected to processes of sociocultural and natural evolution. However, affordances and ecological information would not be pregiven or ready-made since they only emerge when the proper conditions for their enactment occur.

My argument in favor of an enactive account of affordances goes as follows. First, I expose the dynamical definition of Chemero that gets the work of Gibson closer enough to the vision of the enactive approach to see that it is possible to understand affordances as enacted instead of pregiven (Sect. 5.4.1). The fundamental argument of affordances as subject-independent is the claim that affordances are perceived without mental representations thanks to the existence of ecological information. I will argue that ecological information is located in the environment, but it is enacted as ancestral layers of more basic affordances that have become invisible as affordances because they are incorporated into the meaningful environment of an agent. This ancestral layer of affordances serves as scaffolding for the enactment of new layers of affordances. I explain the relation between ancestral and new layers of affordances with the phenomenological concept of ex-corporations (Sect. 5.4.2). Finally, I present the ecological and dynamical nature of the body–world entanglement, implied by the enactive account of affordances, with Merleau-Ponty's notion of spatial levels (Sect. 5.4.3).

5.4.1 *Chemero's Dynamical Account of Affordances*

Anthony Chemero has given one of the most influential interpretations of Gibson's theory of affordances, which he rightly conceives as relational. He does not see affordances as relations between the organism's intrinsic aspects and the environment but as relations between situational features and bodily skills (Chemero, 2009). This redefinition of affordances is welcome, as it gives this concept a dynamical character it previously lacked. Moreover, it facilitates our understanding of the central role of development and the concrete interaction between an agent and the environment to enact affordances.

Chemero first tells us that what matters for an agent to perceive an affordance is not merely to have the appropriate physical body to accomplish an action but to have the *ability* or the *skill* to perform such an action. For instance, I have never learned to play soccer as many friends have. Once, two of my friends and I were walking in the park, and when a soccer ball appeared, my friends immediately started playing with the ball. They passed the ball one each other, walked and dribbled with the ball with their feet, and finally kicked the ball back to its owners. They perceived many affordances in the ball that I could not, given that, if I tried, I could barely kick the ball in a straight line. Therefore, it does not matter if I have the same type of body as my friends do. As long as I do not possess the necessary skills, I cannot perceive what the ball affords.

Our bodily skills change all the time, from the moment we are born (and maybe before) to the moment of our death. Hence, our perception of affordances is constantly changing, and for Chemero, this is what Cesari et al. (2003) experiments prove. In these experiments, Cesari and colleagues demonstrated that people change their perception of the climb-ability of stairs as they lose their climbing skills due to aging factors. A high stair, for instance, was perceived as climbable by young people but not so by older people (Chemero, 2009).

Therefore, it is not enough to have a particular type of body or for there to be specific characteristics in the environment to perceive an affordance. An agent must have the necessary skills to complete an action to perceive the correspondent affordances. Changes in the body occur all the time, either because we get injured or because we lose the sensorimotor habits of a practice, resulting in the inability to recognize affordances. We can also start perceiving new affordances by acquiring new body skills or incorporating artifacts into our natural bodies. The cane of a blind person, the car of a driver, the boots of an alpinist, these things and others can change the layout of affordances (Chemero, 2009).

The ecological realm, for Chemero, is, therefore, a highly dynamic field that constantly changes at multiple temporal scales (see also Gastelum, 2020). At the behavioral time, affordances are differently perceived if we are tired or full of energy. In developmental time, we can change according to fluctuations in our sensorimotor repertoire, our use of artifacts, or changes in our physiology. At the evolutionary time, changes in our genomic and extra genomic heritage can also transform the available affordances of our species.

Material changes in the environment also alter the layout of affordances, destroying and creating new possibilities for action that can change our behavioral, developmental, and phylogenetic paths. On the side of the environment, however, many other dynamic and relational aspects can contribute to the perception of affordances.

For Chemero, our perception of affordances depends more directly on the situational context of the environment than it does on the properties of the objects that comprise the environment. In the example I gave earlier, my friends passed each other the ball so that they perceived in the ball an opportunity to kick it and pass it. These affordances are not intrinsic to the roundness, weight, softness, etc., of the ball. There is a context where the ball exhibits its kick-ability and pass-ability; the flat floor of the park, the ball approaching the leg with a certain velocity, another person able to receive the ball, as well as the fact of being in a park on a sunny day in Montreal. What is more concrete and directly perceived by my friends is something like "it is the right time to kick the ball and play" more than any judgment about whether the ball has the characteristics necessary for playing.

This redefinition of the environment can transform our conception of affordances and the environment where affordances appear, probably more than has been recognized by other followers of Gibson. Chemero's description of the environment as an ecological field is close to Merleau-Ponty's (2012) description of the phenomenal field.

Merleau-Ponty's notion of field comes from Gestalt psychologists, who borrow this term from physics. A standard definition of a field in physics describes an abstract space containing an *x* number of quantitative values or vectors representing a series of concrete forces at play in such a spatial domain. In perception, we also face a spatial domain where different forces can attract our attention, afford possibilities for action, constrain our actions, or solicit a determinate action from us. The classic example of this is a soccer field that appears for a player as a field of forces that constantly change. The limits of the field, the other players, and the ball are just the most evident forces that constitute the lived soccer field of a player (Merleau-Ponty, 1942, pp. 168–169).

Moreover, Chemero's description tacitly introduces the idea that many aspects of the sort that Husserl would call horizonal are needed to determine the enactment of an affordance. Following Husserl, a horizon is something that co-determines the objects that our consciousness intends (Sect. 4.3.1). In the above example, the park, the sunny day, the bodily skills, and the ball's velocity are all part of the co-determinations of ball affordances. The same ball in a sports shop or in the hands of a child does not reveal the same affordances.

Suppose Chemero's interpretation of Gibson's theory of affordances is correct. In that case, the ecological dimension of cognition is a field we cannot entirely determine before the context of action is constituted. For this reason, it follows that *it is in action itself that affordances are enacted* and then appear as correlations of motor habits. These correlations are nonetheless enacted in a broader context that evokes the bodily action, bringing forth the required affordances of the current situation (e.g., the demand for kicking the ball). For all that, affordances are not the sheer mirrors of these habits. Although enacted in the history of the agent–environment system, affordances become sedimented in the environment. They complement and interact with the body, creating tensions and disparities that motivate the self-transformation of the body–world entanglement.

5.4.2 *Ex-Corporations: The Horizons of the Ecological Field*

Chapter 4 argued that Merleau-Ponty's notion of the body schema cannot be reduced to the self-organization of the body in relation to a bodily task (Gallagher, 2005). We should rather conceive the body schema as implying the body–world entanglement. The point here is that the unity of the body and the world precedes any distinction we can make of the agent and the environment.

Phenomenologically, the bond that unites the body and the world is often understood as a relation of co-determination: my motor ability for climbing stairs co-determines the climb-ability I perceive of these stairs. In turn, the motor action I need for climbing stairs co-determines how stairs afford my body the possibility of climbing them. Nevertheless, the body–world entanglement is more than a co-determination of the body and the world because neither simply mirrors each other. There is a constant possibility of discrepancy between what my body aims to do and how the world responds to these aims. I have already mentioned these discrepancies, describing how materials resist flowing smoothly with the habitual actions of the body (Sect. 5.2.2). On some occasions, body–world discrepancies become so insignificant that some aspects of the environment can become incorporated as part of the self and cease being lived as part of the environment. I described this phenomenon as bodily incorporations (Sect. 5.1.2).

Although we can incorporate many features of the environment, the world is inexhaustible, and I cannot expect to become one with it. The gap between ourselves and the world, between us and the other, between us and ourselves, is intrinsic to our cognitive lives (Merleau-Ponty, 1964, 2012). There is always something that cuts across the flux of my actions and aims, just as there is always something unexpected and unintended that becomes part of our lives at every moment (Dastur, 2000). The disparity between the body and the world means that we continuously update our acquired habits and perceptions, thus reconfiguring our normative domain of interactions with the environment. When new norms or new normative domains are enacted, it is not unusual that the older norms stay present in the lived world, even if they become increasingly unnoticed.

When my friends learned to play soccer, they already had a repertoire of bodily skills. They knew how to walk, run, and jump. The different ways to handle the ball in the game became an additional set of norms to the habitual sensorimotor repertoire of my friends. Now, when they play, they no longer need to worry about how to run or walk. Their attention is on the ball and the situation on the field. Playing soccer, nonetheless, would be impossible without the previous existence of more basic sensorimotor norms.

The normative background of playing soccer is not reducible to the habits of the body. Such a background is also sedimented in anchorage points of the field that my friends use as orientation points for their practice (Talero, 2017). When attacking, soccer players run toward the goal of their contenders, and their movements are intrinsically oriented to putting the ball in that goal. The field thus appears as oriented, as structured with positive and negative forces for the players' actions. Likewise, when I walk from the metro to my house, I do not ask myself if the floor is flat and solid, if I need to take the stairs, or if I should turn left or right after the exit. In the bodily practices we master, *many aspects of the environment become invisible to our sight*, but they are nonetheless constitutive of our actions and perceptions. Without the many overlooked aspects of the visual field, it would be impossible for me to know where my house is or for the soccer players where the other team's goal is.

Morris (2004) calls *ex-corporations* all those aspects topologically located on the side of the environment, but that remain in the constitutive background of our lived experiences. Contrary to incorporations, which are portable aspects of the environment (e.g., the cane of a blind person), ex-corporations are not portable and remain situated in places (e.g., the door frame of my bedroom). As a result, they appear to be subject-independent but are not. They are the counterpart of bodily habits, or, better yet, bodily habits are the counterpart of the places a body inhabits (Casey, 1984).

Ex-corporations are not intrinsic features of the environment; they are related to my bodily habits but located outside the body's boundaries, so they are not incorporated but *excorporated*. They are no less relevant for our actions and experiences than those objects that we bodily incorporate. In the same way that a blind person can no longer perceive a distant object without her cane, I would not know where my house is if I could not see the pharmacy and the hospital outside the subway. Indeed, when there is a large amount of snow, the sidewalks vanish from our sight, and we can quickly lose our sense of where the road ends, and the sidewalk begins.

Ecological information is not different from these anchorage points. The lawful correlation of sensing and movement when I walk to the coffee shop counter, for instance, depends on basic sensorimotor coordination of the body–brain–environment system (Di Paolo et al., 2017), making possible the perception of affordances such as the availability of the cup of coffee for being reachable. However, we must acknowledge that its function as vectors of orientation is not determined by an intrinsic correlation between movement and sensory feedback or between the properties of the environment and the properties of the body. All these aspects matter for the constitution of the ecological field, but they are not enough. Ecological information depends on the active bodily engagement of an agent in a situational context, and it is in the constitution of the situation that we must look for the ultimate source of the norms of sense-making. Chapter 6 will address the situated normativity of action and perception. In what follows, I shall describe, from a phenomenological perspective, the origin of ecological information as spatial levels and show how the shift of these levels describes the self-transformation of the body–world entanglement.

5.4.3 Spatial Levels and the Self-Transformation of the Body–World Entanglement

Merleau-Ponty's *Phenomenology of Perception* has been one of the most influential philosophical works of embodied cognition. Nevertheless, little attention has been paid to the concept of spatial levels that has more recently emerged as relevant for the contemporary ontological interpretations of Merleau-Ponty's phenomenology (e.g., Marratto, 2012; Morris, 2018). For many reasons, the description of levels (*niveaux*) is also suitable for our discussion about the ecological dimension of norm development.

First, the concept of spatial levels designates the implicit normativity at play in ecological fields since these refer to the anchorage points that orient our bodily actions. Second, these levels are not subject-independent like ecological information. On the contrary, they imply the entanglement of the body and the world. The normativity of levels depends on the interactional habits incarnated in the body and sedimented in the environment. Third, levels are temporary open-ended structures that can evolve and shift from one level to another, so they also describe the development of norms in the interactions of an agent–environment system. Finally, since levels denote the sedimentation of the norms at play in the agent–environment system, on the side of the environment, they can create tensions with bodily habits that propel the agent–environment system to its self-transformation. In short, the concept of levels can bring together the claims of the enactive approach, ecological psychology, and phenomenology around a definition of sense-making as norm development.

In his *Phenomenology of Perception* (2012), Merleau-Ponty describes the general notion of space from a phenomenological standpoint, which includes the body as a constitutive part. Merleau-Ponty highlights that our experience of space usually implies a particular orientation (e.g., up, down, left, right). This spatial orientation is given to us without the need of our conscious reflection. The primordial sense of space is not the spatiality described by the abstract geometrical dimension of Newtonian physics, a form of space that works as a sort of container for the objects and events in the world (Casey, 1996). Merleau-Ponty (2012) calls this conception *positional spatiality*. For Merleau-Ponty, the primordial form of spatiality is a *situational spatiality*. It involves the active engagement of the body in accomplishing motor tasks and the horizonal domain of all our possible bodily actions. Therefore, the orientation of space we directly perceive is not grounded on something external to the body–world link that motor intentionality establishes (Sect. 4.3.2). Instead, it is a more profound manifestation of this primordial relation.

In this context, the more concrete or delimited regions of space that we call *places* (Casey, 1993) offer *anchorage points* that allow our bodies to situate themselves (Merleau-Ponty, 2012). The anchorage points give places stability, instituting what Merleau-Ponty called *spatial levels* (Merleau-Ponty, 2011, 2012). These levels are norms of perception because they point to the optimal ways our body interacts with the environment (Talero, 2005). Namely, the lived orientation of a spatial level implies the anticipations or expectations of our motor actions. These expectations can be satisfactorily fulfilled by the current conditions of the environment that show at once further possibilities to explore and manipulate this environment. All these happen according to the norm of the perceived level. Perception of places, like the perception of things, involves perceptual norms.

Unlike things, however, places are not usually the focus of our attention; instead, they serve as stable backgrounds for our everyday activities. They form, one could say, the horizonal aspect of our perceptual intentions. The ubiquitous presence of some spatial levels (the more general ones) requires that we alter the ordinary conditions of our sensorimotor interactions to recognize them. This is what Merleau-Ponty did through his interpretation of some classical experiments.

Merleau-Ponty (2011, 2012) refers, for instance, to Stratton's experiment where a subject uses goggles that inverse the visual field for eight days. The visual field is perceived upside down at the beginning. After a couple of days of use, the subject starts to see normally but starts feeling that her body is inverted. After eight days of use, the whole sensorimotor interaction is finally readapted, and the new visual field appears normally. Merleau-Ponty also discusses Wertheimer's experiment, where a subject is put in a room, where she can perceive only through a mirror that distorts what she sees; in optical terms, what she sees deviates by 45 degrees from vertical. The subject initially sees everything obliquely, and even the movement of objects in the visual field is perceived with a similar deviation. However, after a few minutes, the subject perceives the visual scene vertically again.

These examples allowed Merleau-Ponty to argue that some spatial level exists (according to which things seen count as vertical or deviated from verticality) in our habitual sensorimotor interactions that can be, nonetheless, altered if we modify the usual sensorimotor correlations. However, these experiments also show that new spatial levels can be established despite these changes, and, more importantly, these changes are not arbitrary. The anchorage points and the whole structure of the original level are transposed into the new level. Consequently, it seems that the visual structure, constituted by levels grounded on sensorimotor interactions, is a formal structure transposable into another level structure, just as when we transpose a melody from one tonality into another (Merleau-Ponty, 1942, p. 87).

If this interpretation is correct, perceptual norms are not intrinsic to the specific conditions of sensorimotor correlations, as ecological laws predict. The sensorimotor loop is altered, but the same form is perceived. From a phenomenological standpoint, we can explain that our perceptions are based on anticipations of potentialities for motor actions. However, anticipations are themselves grounded on the sensorimotor habits previously acquired by the perceiver. Namely, bodily habits constitute the body schema (Merleau-Ponty, 2012). These habits are correlated to the motor significations perceived in the environment as anticipations and motivations for motor action. Therefore, when a level is forced to shift into another level by changes induced in the sensorimotor loops, the body tries to use its habitual sensorimotor coordination but is forced to reorganize it according to the new circumstances. Since finding similar anchorage points in the emergent sensorimotor dynamic is possible, the habitual form is transposable into a new level.

Merleau-Ponty offers a less dramatic level shift by describing what happens when an organist plays on a new keyboard (Merleau-Ponty, 2012, pp. 146–147). The structure of the new keyboard makes sense thanks to the interactional structure previously acquired with the habitual keyboard.

Initially, the disparities between the anticipations of the body and the current conditions of the environment provoke readjustments in the action–perception cycle. However, after an hour or so, the organist assimilates the new conditions, establishing a new level.

One of the crucial aspects of this description of spatial levels is that perceptual norms have an open-ended character, exhibited by the examples above. This open-ended character of norms also seems necessary to explain why the interactions between the body and the environment remain open to readjustments that follow predictable paths inherent to the normative frameworks (levels) previously enacted. The description of levels is ultimately a description of an endogenous developmental process of the sensorimotor coupling of an agent and the environment. This process entails reconfiguring the already existent normativities that result from resolving a conflict between the body and the world, having both a normative status. The new keyboard is not a senseless aspect of the environment for the piano master; it is a level that has a non-coincidence with her bodily habits. It is a level because the pianist's hands recognize the piano structure. However, the body needs to alter its sensorimotor norms to grasp it more thoroughly and make sense more adequately, as in Stratton's experiments.

Spatial levels can offer a new path for interpreting Gibson's notion of ecological information. The invariants that structure the visual field are ultimately the anchorage points of a level that enables the perception of more complex affordances. Ecological information does not consist of intrinsic structures or mere correlations between raw movement and sensory feedback. Instead, these structures are more primitive forms of affordances correlated to basic motor skills. This is like acquiring new bodily skills, as when we begin to play soccer, based on older skills such as running and kicking.

The notion of spatial levels also reveals that the past of the body–world entanglement constitutes every cognitive act. The past becomes invisible in our sensorimotor habits (body schema) and in the anchorage points of orientation that constitute the sense of an experience (Merleau-Ponty, 2011, pp. 143–144). This invisible past becomes visible in the perceptual field as a resumption (*reprise*) or a reenactment born in the disparity between the habitual and the unexpected.

This role of the past in the constitution of sense-making norms can raise some further questions about its influence on the constitution of a concrete autonomous system. Such a question should lead us to embed the dynamics of the development of individuals in a dynamic of transgenerational lineages and a natural and cultural history that transcends any singular individuality.

Chapter 6 delves deeper into how spatial levels imply norms of places and how the development of these norms more properly describes sense-making. This will lead me to consider the embedded nature of agents in their social and natural history, one that is always shared with other agents and constructs the tangled web of the world we inhabit.

Notes

1. The enactive approach distinguishes between two types of incorporations: tool-incorporation and mutual incorporation (Fuchs & De Jaegher, 2009). Tool-incorporation designates the sort of incorporation I have just described. Mutual incorporation rather refers to the way that two agents enact jointly a new meaning. In this sense, dancing in couple involves a mutual incorporation. The movements of each dancer make sense only insofar as the other is actually or virtually present. The smooth coping of a good dancing performance can get closer the phenomenon of mutual incorporation to the one of tool-incorporation. However, according to the dynamical model of the enactive approach, the sort of tension at play in an interaction between two autonomous systems and between an agent and a tool is not the same. The coordination between two agents is always more conflictive and harder to reach, although it could be easily to handle by a mutual coregulation. The feeling and effect of mutual co-regulation is precisely what is always missing in tool-incorporation.
2. The notion of "information pickup," as well as others in Gibson's work (e.g., ecological information, affordances), is controversial an open to multiple interpretations. I am referring here to the most common and maybe superficial understanding of "information pickup," which consists of the description of how perceivers obtain information from the visual flow thanks to their own displacements in space. This is exemplified in Gibson's (1966) analysis of pilots handling information for landing a plane. Such analyses acknowledge the importance of self-movement for obtaining information in the flow but neglect the central role that self-organization processes of the body play to constitute as meaningful the information of the flow (e.g., the bodily skills required for piloting a plane). The difference between bare movements in space and self-organized movements of the body implicated in the acquisition of a body skill is what above Bredlau's example illustrates. The same distinction is made by scholars of the enactive approach in their differentiation between a sensorimotor environment and sensorimotor habits.

References

Baber, C., Chemero, T., & Hall, J. (2019). What the jeweller's hand tells the jeweller's brain: Tool use, creativity and embodied cognition. *Philosophy & Technology, 32*(2), 283–302.

Bach-y-Rita, P., & Kercel, S. W. (2003). Sensory substitution and the human–machine interface. *Trends in Cognitive Sciences, 7*(12), 541–546. https://doi.org/10.1016/j.tics.2003.10.013

Bateson, G. (1987). *Steps to an ecology of mind: Collected essays in anthropology, psychiatry, evolution, and epistemology.* Jason Aronson Inc.

Bredlau, S. (2006). *Learning to see: Merleau-Ponty and the navigation of "terrains"* (Vol. 8, pp. 191–198). Chiasmi International.

Casey, E. S. (1984). Habitual body and memory in Merleau-Ponty. *Man and World Man and World, 17*(3–4), 279–297.

Casey, E. S. (1993). *Getting Back into place: Toward a renewed understanding of the place-world.* Indiana University Press.

Casey, E. S. (1996). How to get from space to place in a fairly short stretch of time: Phenomenological prolegomena. In S. Feld & K. Basso (Eds.), *Senses of place* (pp. 13–52). School of American Reserch.

Cesari, P., Formenti, F., & Olivato, P. (2003). A common perceptual parameter for stair climbing for children, young and old adults. *Human Movement Science, 22*(1), 111–124. https://doi.org/10.1016/S0167-9457(03)00003-4

Chemero, A. (2009). *Radical embodied cognitive science.* MIT Press.

Clark, A. (2008). Pressing the flesh: A tension in the study of the embodied, embedded mind? *Philosophy and Phenomenological Research, 76*(1), 37–59. https://doi.org/10.1111/j.1933-1592.2007.00114.x

Dastur, F. (2000). Phenomenology of the event: Waiting and surprise. *Hypatia, 15*(4), 178–189.

De Jaegher, H. (2018). The intersubjective turn. In L. de Bruin, A. Newen, & S. Gallagher (Eds.), *The Oxford handbook of 4E cognition.* Oxford University Press.

De Jaegher, H., & Di Paolo, E. A. (2007). Participatory sense-making. *Phenomenology and the Cognitive Sciences, 6*(4), 485–507. https://doi.org/10.1007/s11097-007-9076-9

Di Paolo, E. A. (2009). Extended life. *Topoi, 28*(1), 9.

Di Paolo, E. A. (2016). Across the Uncanny Valley: The ecological, the enactive, and the strangely familiar. *Constructivist Foundations, 11*, 327–329.

Di Paolo, E. A., Barandiaran, X. E., Beaton, M., & Buhrmann, T. (2014). Learning to perceive in the sensorimotor approach: Piaget's theory of equilibration interpreted dynamically. *Frontiers in Human Neuroscience, 8*, 551.

Di Paolo, E. A., Buhrmann, T., & Barandiaran, X. E. (2017). *Sensorimotor Life: An Enactive Proposal.* Oxford University Press.

Di Paolo, E. A., Cuffari, E., & De Jaegher, H. (2018). *Linguistic bodies: The continuity between life and language.* MIT Press.

Di Paolo, E. A., & Thompson, E. (2014). The enactive approach. In L. Shapiro (Ed.), *The Routledge handbook of embodied cognition* (pp. 86–96). Routledge.

Fuchs, T., & De Jaegher, H. (2009). Enactive intersubjectivity: Participatory sense-making and mutual incorporation. *Phenomenology and the Cognitive Sciences, 8*(4), 465–486.

Gallagher, S. (2005). *How the body shapes the mind.* Oxford University Press.

Gastelum, M. (2020). Scale matters: Temporality in the perception of affordances. *Frontiers in Psychology, 11*, 1188. https://doi.org/10.3389/fpsyg.2020.01188

Gibson, J. J. (1966). *The senses considered as perceptual systems.* George Allen & Unwin Ltd.

Gibson, J. J. (2002). A theory of direct visual perception. In A. Noë & E. Thompson (Eds.), *Vision and mind: Selected readings in the philosophy of perception* (pp. 77–90). MIT Press.

Gibson, J. J. (2015). *The ecological approach to visual perception.* Psychology Press.

Heft, H. (2015). *Ecological psychology in context: James Gibson, Roger barker, and the legacy of William James's radical empiricism.* Psychology Press.

Held, R., & Hein, A. (1963). Movement-produced stimulation in the development of visually guided behavior. *Journal of Comparative and Physiological Psychology, 56*(5), 872.

Heras-Escribano, M. (2019). *The philosophy of affordances.* Springer International Publishing.

Heras-Escribano, M., & de Pinedo, M. (2016). Are affordances normative? *Phenomenology and the Cognitive Sciences, 15*(4), 565–589.

Hurley, S., & Noë, A. (2003). Neural plasticity and consciousness. *Biology and Philosophy, 18*(1), 131–168.

Hutto, D. D., & Myin, E. (2017). *Evolving enactivism: Basic minds meet content.* MIT Press.

Malpas, J. E. (2004). *Place and experience: A philosophical topography.* Cambridge University Press.

Marratto, S. L. (2012). *The intercorporeal self: Merleau-Ponty on subjectivity (SUNY series in contemporary French thought).* State University of New York Press.

Merleau-Ponty, M. (1942). *The structure of behaviour.* (A. L. Fisher, Trans.). Duquesne University Press.

Merleau-Ponty, M. (1964). *Le visible et l'invisible: suivi de notes de travail. par Maurice Merleau-Pont.* Texte établi par Claude Lefort. Gallimard.

Merleau-Ponty, M. (2011). In E. de Saint Aubert & S. Kristensen (Eds.), *Le monde sensible et le monde de l'expression: cours au Collège de France: Notes, 1953.* MetisPresses.

Merleau-Ponty, M. (2012). *Phenomenology of perception.* (D. Landes, Trans.). Routledge.

Montuori, A. (2003). The complexity of improvisation and the improvisation of complexity: Social science, art and creativity. *Human Relations, 56*(2), 237–255.

Morris, D. (2004). *The sense of space*. State University of New York Press.

Morris, D. (2018). *Merleau-Ponty's developmental ontology*. Northwestern University Press.

Mossio, M., & Taraborelli, D. (2008). Action-dependent perceptual invariants: From ecological to sensorimotor approaches. *Consciousness and Cognition, 17*(4), 1324–1340.

O'Regan, J. K., & Noë, A. (2001). A sensorimotor account of vision and visual consciousness. *Behavioral and Brain Sciences, 24*(5), 939–973.

Ramstead, M. J. D., Veissière, S. P., & Kirmayer, L. J. (2016). Cultural affordances: Scaffolding local worlds through shared intentionality and regimes of attention. *Frontiers in Psychology, 7*, 1090.

Reed, E. S. (1996). *Encountering the world: Toward an ecological psychology*. Oxford University Press.

Rietveld, E. (2008). Situated normativity: The normative aspect of embodied cognition in unreflective action. *Mind, 117*(468), 973–1001.

Rietveld, E., & Kiverstein, J. D. (2014). A rich landscape of affordances. *Ecological Psychology, 26*(4), 325–352.

Sawyer, R. K. (1992). Improvisational creativity: An analysis of jazz performance. *Creativity Research Journal, 5*(3), 253–263.

Schiavio, A., & De Jaegher, H. (2017). Participatory sense-making in joint musical practices. In M. Lessaffre, P.-J. Maes, & M. Leman (Eds.), *The Routledge companion to embodied music interaction* (pp. 31–39). Routledge.

Sepúlveda-Pedro, M. A. (2020). Levels and norm-development: A phenomenological approach to enactive-ecological norms of action and perception. *Frontiers in Psychology, 11*, 1666. https://doi.org/10.3389/fpsyg.2020.01666

Talero, M. (2005). Perception, normativity, and selfhood in Merleau-Ponty: The spatial 'level'and existential space. *The Southern Journal of Philosophy, 43*(3), 443–461.

Talero, M. (2017). The image and the work space: Merleau-Ponty and Levinas on passivity and Rythmic subjectivity. In K. Jacobson & J. E. Russon (Eds.), *Perception and its development in Merleau-Ponty's phenomenology* (pp. 40–50). University of Toronto Press.

Thompson, E., & Stapleton, M. (2009). Making sense of sense-making: Reflections on enactive and extended mind theories. *Topoi, 28*(1), 23–30.

Turvey, M. T. (1992). Affordances and prospective control: An outline of the ontology. *Ecological Psychology, 4*(3), 173–187. https://doi.org/10.1207/s15326969eco0403_3

Turvey, M. T., & Carello, C. (1995). Some dynamical themes in perception and action. In R. F. Port & T. van Gelder (Eds.), *Mind as motion: Explorations in the dynamics of cognition* (pp. 373–401). MIT Press.

van Alphen, F. (2014). Tango and enactivism: First steps in exploring the dynamics and experience of interaction. *Integrative Psychological and Behavioral Science, 48*(3), 322–331.

Villagrán, C., Contreras, F., Cornejo, F., Figueroa Cuadra, M., Valenzuela-Bezanilla, D., Luraschi, R., et al. (2020). Understanding gold toxicity in aerobically-grown *Escherichia coli*. *Biological Research, 53*, 1. https://doi.org/10.1186/s40659-020-00292-5

Walton, A. E., Richardson, M. J., Langland-Hassan, P., & Chemero, A. (2015). Improvisation and the self-organization of multiple musical bodies. *Frontiers in Psychology, 6*, 313.

Warren, W. H. (1984). Perceiving affordances: Visual guidance of stair climbing. *Journal of Experimental Psychology: Human Perception and Performance, 10*(5), 683–703. https://doi.org/10.1037/0096-1523.10.5.683

Wittgenstein, L. (2009). *Philosophical investigations*. (G. E. M. Anscombe, P. M. S. Hacker, & J. Schulte, Trans., P. M. S. Hacker & J. Schulte, Eds.) (4th ed.). Wiley-Blackwell.

CHAPTER 6

Sense-Making as Place-Norms: Inhabiting the World with Others

In the previous chapters, I defined sense-making as norm development because it involves reconfiguring the existing normative domain of the body–world entanglement. This entanglement implies that an agent and the environment are never totally alien to each other. Instead, they are always connected by norms of interaction enacted in the past of the agent–environment system. At the same time, we must remember that the agent and the environment are never fully fused or coordinated. There is always a disparity between the already enacted norms and the emergent circumstances that challenge the "adequate" interactions of the system. In this picture, the environment is not a purely physical realm that acquires meaning solely as a result of the activity of agents; instead, it is a normative force that remains in constant tension with them. This tension is what motivates the ongoing self-transformation of the body–world entanglement.

This chapter aims to improve our understanding of the self-transformative movement of the body–world entanglement through enacting norms of sense-making, and it will do so by locating this complex process as emerging from the *situatedness* of agents. I conceptualize this situatedness as an *enactive place*. This concept depicts the metastability of the interactional field that gathers together all the multiple aspects that constitute a given situation. It is within this situatedness—or within the emplacement of agents—that sense-making occurs. Throughout this

M. A. Sepúlveda-Pedro, *Enactive Cognition in Place*, New Directions in Philosophy and Cognitive Science,
https://doi.org/10.1007/978-3-031-20282-7_6

book, I have argued that sense-making cannot be understood simply by looking at an agent's autonomy, and at the material constraints of the environment. Instead, the concept of situated enactive place enhances both our understanding of norm development (Chap. 4) and the role of the environment as an active normative field (Chap. 5) in processes of life and cognition.

For a more precise description of enactive place, this chapter defines the central characteristics of place based on Edward Casey's phenomenologically inspired philosophy (Sect. 6.1). I use the paradigmatic example of Mexican *Chinampas,* or the traditional floating gardens of Mesoamerican agriculture, to show that place as described by Casey involves an enactive constitution. The enactive characterizations of place lead us to conclude that the constitutional force of place is a *situated normativity*. The ecological approach of the skilled intentionality framework provides a good basis for understanding the nature of this situated normativity (Sect. 6.2). However, this concept still needs to be refined, because the skilled intentionality framework neglects the participation of agents in the constitution of situational norms. Hence, I suggest a truly enactive account of situated normativity, one based on the most recent works of the enactive approach. Crucially, the concept of enactive situated normativity recognizes that agents actively participate in the constitution of sociomaterial norms (Sect. 6.3). Since my aim is not to attribute the ecological dimension of sense-making to sociocultural practices alone, but to all other forms of sense-making, I suggest at the end of the chapter that enactive situated normativity *also* exists in natural places we commonly refer to as ecological niches (Sect. 6.4). These niches can entangle human and nonhuman agents within the same ecological–enactive web, thereby continuously evolving as a resilient unity. However, I will argue that this argument does not ignore the significant differences between the sociocultural and the pre-cultural bodily interactions of agents recognized by both the enactive approach, and by phenomenology.

6.1 An Enactive Theory of Place

We have already seen in previous chapters that from the perspective of the enactive approach, our commonsense conceptions of the environment, as a subject-independent domain, are not adequate to describe the proximate surroundings of living agents. We have also seen that a twofold meaning of the term 'environment,' as both *Umwelt* (the subject's relative meaningful world) and *Umgebung* (the physical surroundings), was needed to understand life and cognition from the first and third-person perspectives, respectively. This section introduces another aspect to describe the immediate environment of living agents that I call *enactive place.*

For a basic definition of place as enactive, I appeal to Casey's phenomenological descriptions of place (Sect. 6.1.1). From a phenomenological perspective, these descriptions reveal that place has no absolute foundations. Instead, place is part of the complex normative field that living agents enact in the temporal unfolding of their interactions with their local surroundings. The paradigmatic example of Mexican Chinampas illustrates Casey's description of place and helps us see that place is constituted by complex interactional processes of historical development (Sect. 6.1.2). At the end of this section, I argue that such development involves not only the transformation of matter but also the transformation of normative relations among the multiple inhabitants of place (Sect. 6.1.3).

6.1.1 From Space to Place

In his analysis of the history of Western thought, Casey (1998) distinguished between modern, premodern, and postmodern characterizations of place. From a phenomenological standpoint, Casey argued that the premodern and postmodern characterizations more accurately define the local spaces we inhabit. They can also guide us to a more profound philosophical account of bodily existence than the modern characterization. Here, I will refer to the modern characterization of place as the absolutist theory of place, denoting the premodern and postmodern as part of an enactive theory of place. In doing so, I go beyond Casey while still drawing from his thought, pointing to significant resonances with issues of enaction that I have traced from the beginning of this book.

These two theories, absolutist versus enactive theories of place, are in radical opposition. Whereas the absolutist view assumes that a general and highly abstract notion of space is a precondition (substantial or formal) for the existence of place, the enactive theory holds that place has primacy over any general conception of space (and time). The absolutist theory of place tends to coincide with the usual views of scientists and philosophers, inasmuch as they adopt what Husserl called the natural and the naturalistic attitudes of experience. However, an enactive theory of place coincides with definitions proper to a phenomenological (although not necessarily a transcendental) attitude.

In our natural attitude, we usually talk about places as delimited spatial areas where objects, people, and narrower spatial areas are localized and contained. In each room of my apartment, for instance, things and people occupy a spatial area. However, these rooms are contained within the outer spatial boundaries of the apartment (i.e., the walls, the floor, windows, and the ceiling). The apartment is at once part of a building located in a block of buildings. We may continue to describe wider spatial regions that contain this block, this neighborhood, this city, or even the entire planet earth. If we think more widely, we can imagine that a bigger spatial container for our planet is needed. We may therefore conclude that the universe is the largest spatial container for any physical object (Casey, 1996).

Seen from the point of view of the naturalistic attitude of modern science, place, as a subregion of space, is less concrete than a spatial container, while at the same time founded on the abstract characterization of space as an absolute dimension. For instance, Newton's (1999) classical physics describes space and time as absolute dimensions where all physical events, such as the motion of substantial objects, occur. Space, for Newton, is just the setting or the background of all physical events, but one that remains independent of these phenomena. Therefore, in this context, place is a specific region of absolute space and seems more concrete because it contains both objects and events. Still, as a mere spatial portion of an absolute dimension, places remain pieces of the abstract void (Casey, 1998, p.147).

Newton's characterization of space and time as absolute dimensions prompted Kant to understand that space and time are not phenomena that we can experience but the conditions of possibility of our experience. That is, space and time are transcendental manifolds of intuition instead of dimensions of nature as such (Kant, 1998). As subjective categories, space and time work once again as absolute foundations, but this time for experience and consciousness, for possessing a world of sense and meaning. For both Newton and Kant, space and time are the ultimate and absolute foundations of the world of nature. They logically precede any concrete event or the existence of any physical object we perceive. Therefore, place is secondary and only possible thanks to the abstractions of space and time (Casey, 1998).

As we have seen in previous chapters, both the enactive approach and phenomenology have rejected the claims of modern philosophers and scientists that see cognition and consciousness as having absolute foundations, like that of the subject-independent world or, similarly, the transcendental constitutive mind. These lines of thought hypothesize that either substances or formal structures ground the appearance of any phenomena we experience. However, we have seen that such hypotheses are often based on mere theoretical constructions and naïve presuppositions rather than on observation and analysis of what experience shows.

Like the anthropologist and the phenomenologist, the enactivist should become immersed in the world to disclose how things show themselves to our different modes of experience. This is contrary to the methodology of the classical sciences of nature that isolate phenomena in a lab or in abstract scenarios where things show themselves differently. Although these scientific methods can be helpful for specific purposes, the same methods can obscure our comprehension of phenomena such as place. From our alternative perspective, we can follow Casey's conclusion that place is primordial and that it existentially precedes the general regions of space and time (Casey, 1996). Section 6.1.2 surveys the most relevant characteristics of place and its constitutive norms.

6.1.2 An Enactive Description of Place

It is not easy to define place because place, from this perspective, implies a complex interrelation between parts and wholes, or between the objects situated in place and the place where these objects are situated (Malpas, 2004). This prevents us from carrying out a clear-cut differentiation between matter and form, two aspects traditionally held as separate in conceptual analyses but that converge in the concreteness of place. Therefore, from a phenomenological perspective, one of the main characteristics of place is that place is a unity or a whole that we cannot define independently of the parts that constitute it (Morris, 2018), like in the interpretation of place as a container of things. At the same time, however, the definition of place cannot be arrived at by the sum of its parts because its materiality does not exhaust its definition (Casey, 1996). The constitution of a place rather depends on the already established normative field of interactions between multiple agents and the sedimented materiality of the local environment (Casey, 1998, pp. 206–207). For example, look at the traditional raised fields in the Valley of Mexico called *Chinampas.*[1]

Chinampas are beds of earth and mold surrounded by hedges and sticks. They are part of a technique of Mesoamerican agriculture to form solid surfaces above the water of lakes and marshes (Frederick, 2007).[2] Although these surfaces were made initially to cultivate crops and flowers, they have also served as grounds for houses and buildings (Calnek, 1972; de Ruiz, 1973). A significant aspect of the *chinampas,* for our discussion, is that *chinampas* constitute a highly visible web of ecological interrelations at multiple levels of agent–environment interactions (Eakin et al., 2019; Gliessman et al., 1981; Torres-Lima et al., 1994).

Chinampas are places that their material constitution cannot fully define. The sum of the land, the human and nonhuman agents, and the material resources provided by nature and culture, both inside and outside the *chinampa,* do not exhaust its definition or fully explain its unity. Instead, we must look to the emergent interdependence of multiple agents in relatively coordinated cycles of action, an interdependence that entangles a community of agents and material resources in a shared history, to define place more properly (Casey, 1996).

The human intentions to build *chinampas* are indispensable in determining the ecological entanglement of *chinampas.* However, other living agents and the sedimented materiality of *chinampas* are also active participants in their ecological constitution. Animals and plants motivate and constrain the action of humans and co-participate in the construction of the material landscape of the *chinampa.* Nonhuman agents not only alter the environment, as when they contribute to the chemical composition of the land, but they also interact in a coordinated way with humans to build the shape of *chinampas.* For example, certain kinds of animals nourish from the *chinampa* crops and provoke farmers or *Chinamperos* to cultivate flowers and crops that attract these animals (Gliessman et al., 1981).

The entanglement of culture (as a human bodily practice) and nature (as the organic life of *chinampas*) is not unfamiliar to the more intuitive perception of *Chinamperos* (Cox et al., 2020). They are highly sensitive to the problems caused by disease and lack of fertility in the soil or the contamination in the lake's waters. This is undoubtedly the case for the few *chinampas* that still float on the lake of Xochimilco, south of Mexico City (Eakin et al., 2019).

A second relevant aspect of place is that it consistently exceeds its boundaries (Casey, 1996). For instance, we can say that the spatial limits of *chinampas* are the canals surrounding them, but processes happening inside these boundaries often depend on processes occurring outside. For instance, as cultivation fields, *chinampas* do not rely directly on rainfall. However, the water from the lakes surrounding chinampas is needed to provide the organic matter and minerals for cultivation (Torres-Lima et al., 1994). The sticks and hedges employed to enclose *chinampas* are also found in the milieu of the lake or within other *chinampas. Chinamperos* also plant a particular type of tree found in the local environment, called the *Ahuejote*, which gives more stability to the ground of the *chinampa*. Therefore, the construction and self-maintenance of *chinampas depend* on the resources found within the *chinampas* and in their local environment. However, these constitutional aspects of *chinampas* are not merely local. Agents that participate in the living processes of *chinampas* arrive seasonally from far distances, like birds migrating from North America in winter. These visitors transform the cycles of interaction within *chinampas* every season and contribute to maintaining the lake's ecological balance (Merlín-Uribe et al., 2013). Therefore, we can say that *chinampas* constitute an *existential unity* because the establishment of its specific way of life entangles participants and material structures in an ecological web (see, e.g., Crossley, 2004; Eakin et al., 2019).

In the same way that the permeability of place allows the local environment to affect it, a place may also contribute to the constitution of the local environment. This back-and-forth entails cycles of reciprocal causality that reveal the relational constitution of one place in relation to others, as Casey (1996) described it from a phenomenological perspective. When Mexicas or Aztecs made use of the construction of *chinampas* to build their entire city, for instance, thousands of houses and prominent pyramidal temples such as the *Templo Mayor* were built on the grounds of *chinampas* (Aveni et al., 1988). This constituted a significant transformation of the regional landscape and the historical development of the *chinampas* environment, ending in the current constitution of Mexico City.

The permeability of place is, therefore, not simply material, as this sketch of the ecological interdependence of *chinampas* and lake reveals. The permeability of place also concerns dynamic and transformative aspects. For instance, the change of one aspect of place can drive a transformational shift of place itself, as seen in the movement of *chinampas* into a whole city. We can say that place exhibits norm development and *norm envelopment* since it reveals the entanglement of different dimensions and levels of normativity that determine the experience, the behavior, and the interaction of agents in places (cf. Morris, 2004). Similarly, the normativity and materiality of place also transcend its immediate temporality (Casey, 1998). The materiality of places constantly changes over time, sometimes due to external aspects of the practice that place envelops.

The severe transformation of the environment in the case of Tenochtitlan caused a further transformation of *chinampas* as fields of cultivation. Farmers cultivated new crops in *chinampas* during the colonial period (Torres-Lima et al., 1994). New tools were introduced for the practice of farming. In the twentieth century, most of the few *chinampas* remaining south of Mexico City resorted to growing flowers since they were economically more valued (Narchi & Cristiani, 2015). Due to the economic pressures and the lake ecosystem degeneration, chinamperos are also increasingly changing their modes of existence. They often become part-time farmers because they must work in urban occupations to maintain a minimal economic income. This fact is breaking the relationship of farmers with the land, and new generations are leaving the *chinampas*, sometimes choosing to build houses there instead of agricultural fields. As a result, there is a new ecological disequilibrium in the lake region (Charli-Joseph et al., 2018).

The transformation of a place such as a *chinampas*, one that yet resists dramatic change at multiple levels of interaction in the environmental region, underlines the *resilience* of places and the constitutive force of their unity (G. Barker, 2017). The supporters of niche construction theory (Laland et al., 2000) have suggested that ecological niches, such as the *chinampas* environment, are the fundamental unit of evolution and not merely genes (Dawkins, 2006) or organisms (Lewontin, 1983). Although this claim may be exaggerated, as it ignores the autonomy of biological processes happening in species, groups, and individual organisms, it rightly points to the entanglement of organisms with the local environment and their co-determination from an external point of view. This causal co-determination of ecological niches resonates with the experiential co-determination of place. Individuals are normatively attached to the active life of their surroundings and to the material structures that constitute this life. They also exhibit a peculiar sensitivity to and a behavioral dependency on the structures of place (Heft, 2018). Place is, therefore, a dynamical unity that is in constant self-transformation, materially and normatively, and it is the result of both internal and external factors.

In sum, places are temporally open-ended units, permeable, and flexible in many ways. However, they incarnate the stability of the bodily practices of our lives as individuals, as cultural groups, and I will argue (Sect. 6.4.1), as natural species. Things in the world can be replaced or disappear, while the identity of places remains, not unlike how our bodies replace their material components while preserving a fundamental kind of organization. As with the case of the living body, the identity of place is not intrinsic to its material structure. Instead, the normative structure makes a place what it is; to fully understand this, we must be aware of the entanglement of situations and place.

6.1.3 Place and Levels of Situated Normativity

In Casey's account of place, illustrated by the example of *chinampas*, we have seen that the unity of a place does not simply depend on criteria we impose from our external observations. Instead, an internal web of ecological relations constitutes a resilient system, one that evolves in relation to internal and external components, as well as to other places that surround and envelop its systemic unity. Earlier, I said that the unity of place is existential because such a unity entangles different levels of normativity in close coordination but also in tension and thereby constitutes a particular way of life. This way of life gives a particular *style* to the locality and its inhabitants (Casey, 1996), not so different from the self-organization of the body that gives a particular style to our movements (Merleau-Ponty, 2012, p. 329). This existential unity of a local environment more properly defines a place and characterizes the whole set of normative constraints and affective motivations that trigger the self-transformation of an agent–environment system. As experienced from the inside (i.e., as an inhabitant of place), the unity of a place is determined by the sense of the situations that occur in place. These situations are not mere events or causal processes; they have a meaning experienced only by those who inhabit a place. Situations and places are deeply entangled (McGann, 2014a, 2014b). Situations significantly determine the being of place, but situations are embodied in place.

Although situations are necessarily emplaced, it does not follow that place and situation can simply be identified with one another. The field of relevant aspects in the environment determined by a situation does not exhaust all the richness of place. Place is sedimented by different levels of normativity enacted in the history of an agent–environment system. Some of these levels may not appear relevant for a current situation and may remain in the background. However, they still function as horizonal aspects of that situation. For instance, a room in the philosophy department where seminars *take place* can also be used for a cocktail party. In this situation, relevant aspects of the room as a classroom disappear, and new aspects appear instead. My behavior and that of my colleagues changes as a result. We may talk and laugh louder, joke, drink wine and beer, and listen to music. I can no longer read or take notes, while discussions of philosophical matters have different nuances and rules. We can say that many norms of action and perception have changed with the shift of the situational context of the room.

Despite the changes of these two situations, there are normative layers of the room that persist and give us perceptual clues about more general aspects of the room. For instance, the flatness of the floor I walk over, the color of the walls, or the illumination provided by the lamps of this room. There are also endless external horizons of the room. For example, I need to climb three flights of stairs to get to the room, regardless of its activity. Even if I do not know the number of the room, or if I do not have a name for the room, I can perceptually recognize the room as the same in different situations thanks to external horizons that envelop it as a place. I believe this constancy of place is due to the different levels of generality of the norms of action and perception (see Sect. 5.4.3). While some norms pertain to more specific aspects of a bodily practice that can change with the situational contexts of places, other norms are more general and give to places their more persistent identity.

There is another aspect of place that exceeds situations. As a metastable structure, place embodies the recurrent dynamics of interaction between agents and the environment. However, this structure is labile and suffers transformations not necessarily determined by agents' purposeful intentions or the metastable structure's recurrent dynamics. It is possible to find in places new configurations due to internal or external disturbances that may transform the bodily practices of agents, along with the interrelation of these agents with the environment. For example, governments and social organizations plan urban development in our contemporary cities, but traditionally this development occurred due to changes in societies and their practices that gave a new shape to places. Here, it is evident that transforming places may lead to developing new practices. For example, the lake environment and the practical knowledge of the people inhabiting the basin of Mexico allowed ancient Mexicas to build an entire city in the middle of a lake (Morehart & Frederick, 2014). The appropriation by Mexicas of this place involved adapting agents to the environment, thereby transforming their way of life.

Consequently, the unity of place is significantly determined by the situations occurring in a place, but it is not exhausted by them. The unity of place has different levels of normativity, and the interweaving between the different levels of normativity designates the concrete unity of place. The actual activity of living agents and their interaction with living and nonliving beings in a particular environment contains all the multiple aspects that constitute a place (Casey, 1996). Therefore, the gathering force of place is expressed in the networks of normativities that constitute a place and entangle multiple horizons of experience in a specific but permeable or partially unbounded location. Given this, we will need to understand the norms that constitute place in more detail.

6.2 Ecological Situated Normativity and Norm Attunement

Roger Barker was an ecological psychologist who studied how social environments influence the behavior of human beings. Barker (R. G. Barker, 1968) and his team studied people's behavior in a small town in the Midwest United States. In this study, the team expected to find regular behavior patterns in individuals across different social places. They were surprised to learn that individuals showed less regularity in their behavior in different places than different people showed in the same place. Different people behaved similarly in the drug store, the park, and the public plaza. On the contrary, the same individuals behaved differently in all these different places. Barker concluded that sociocultural environments were a determining factor in explaining the social behavior of people. Barker called these social places *behavior settings*.

Marek McGann (2014a, 2014b) recognizes there is a relation between the theory of behavior settings and the claims of the enactive approach. He acknowledges, for instance, that behavior settings involve a new set of constraints for the sociocultural practices of individuals, but also that the constitution of places involves the active engagement of the participants. He lacks, however, a deeper examination of the normative field of places and their dynamic self-transformation. This can be better done—I think—thanks to my redefinition of sense-making as norm development, the environment as an active normative field, and a phenomenologically informed description of place such as Casey's.

In this section, I argue, like McGann (2014b), that we can put in terms of what the skilled intentionality framework describes as a situated normativity the relation between patterns of social behavior and the specific characteristics of a material environment (Sect. 6.2.1). This normativity involves explicit and implicit social conventions that rule the interactions among people and between people and the material environment. According to the supporters of this ecological approach, however, agents' bodies become affectively attuned to the sociocultural norms of practices thanks to a "skilled intentionality" (Sect. 6.2.2). I argue this concept neglects the enactive nature of situated norms. Skilled intentionality involves processes of bodily attunement to pre-given ecological norms instead of processes of norm enactment based on the individual development of agents (Sect. 6.2.3). Hence, I conclude we need an alternative enactive account of situated normativity.

6.2.1 Situated Normativity

The skilled intentionality framework coined the notion of situated normativity, based on Wittgenstein's (2007, 2009) analysis of lived experience while subjects perform sociocultural practices. Agents like tailors and architects feel discomfort and discontent if they find the conditions of their practices unsatisfactory. If they have enough expertise, they can be moved to take a specific action that improves these conditions. This experience of discomfort and the action itself can happen with or without the agent's conscious reflection (Rietveld, 2008). Skillful agent's sensitivity to the contextual demands of the practice reveals that their bodies are already attuned to the normative framework of that practice. For the skilled intentionality framework and Wittgenstein, this framework is not individual or private but social and public (Rietveld et al., 2018). Sociocultural practices have standards explicitly or tacitly accepted by the community to which participants of these practices belong. Thus, feelings of dissatisfaction and solicitations of action are grounded on public measures. This analysis led scholars adhering to the skilled intentionality framework to adopt the idea that situated normativity does not refer to norms enacted by individuals but to norms that rule the habitual patterns of bodily practices of a sociocultural group. After Wittgenstein, these patterns were called a *form of life* (Rietveld & Kiverstein, 2014).

A form of life is entangled with material structures or physical constraints that help to constitute and shape the normative character of certain practices. For this reason, the scholars of the skilled intentionality framework designate the environmental conditions of human practices as a *sociomaterial environment* (van Dijk & Rietveld, 2017). The paradigmatic case of a form of life is a human sociocultural group, but the notion of a form of life is not exclusive to human beings. Human and nonhuman animal forms of life inhabit spatial regions called *ecological niches*. Again, these niches do not simply refer to the raw material composition of a spatial location but to the entanglement of the materiality and the form of life that inhabit such location. The skilled intentionality framework names this niche a *landscape of affordances* (Rietveld & Kiverstein, 2014).

The skilled intentionality framework distinguishes between two different sets of affordances: the first being a *landscape*, and the second, a *field of affordances* (Rietveld & Kiverstein, 2014). Whereas a landscape of affordances represents all those affordances available to a given form of life, a field of affordances refers to a subset of this landscape, one composed of affordances deemed relevant to a skillful agent. As Chemero (2009) has argued, it is not enough to have a particular type of body to perceive affordances; an agent also needs to possess the appropriate bodily *skills* to exploit these affordances just to perceive them. Likewise, the practice context makes us see some affordances and not others, thereby setting forth the actions required for the concrete happening of a situation. A field of affordances is thus a domain (or set) of *solicitations* that move an agent to act in a bodily practice (Rietveld et al., 2018).

In sum, whereas a field of affordances is concrete and subject-relative, a landscape is abstract and subject-independent. In the case of nonhuman animals, an ecological niche is a relation between the patterns of behavior of a species and the material conditions of their environment. In the case of humans, the ecological niche is the relation of patterns of behavior of a sociocultural group embedded in a material environment. In both cases, the normative framework of practices (the landscape) is defined in reference to a group of individuals and not to individuals themselves; these only become attuned to the pre-given frameworks, thereby perceiving a field of affordances in the environment around them.

We can now better appreciate the relation between the concepts of behavior settings and situated normativity. The ecological niche of a natural form of life, including us, is tantamount to all the potential affordances that our natural species can exploit. As a sociocultural domain, a behavior setting constrains the number of available affordances for human agents because the situated normativity of that setting imposes a new layer of constraints on the affordances of the natural environment. Following the community's standards, agents limit the number of their possible actions and the correspondent affordances (Bruineberg et al., 2018). In the example of *chinampas*, we see it affords multiple actions for human and nonhuman animals. As a behavior setting, *chinampas* have a more reduced number of affordances available than as an ecological niche, for example, inhabiting instead of cultivating *chinampas*.

6.2.2 *Skilled Intentionality*

The skilled intentionality framework explains the acquisition of norms of action and perception by individual agents as the result of a process of attunement with the environment. This is called *skilled intentionality* (Bruineberg & Rietveld, 2014). This concept has two primary sources for its definition: Merleau-Ponty's (2012) phenomenological description of the maximum or *optimal grip* and Friston's (2010) theory of the free-energy principle.

Phenomenologically, skilled intentionality is the tendency of the body and the environment to reach for an optimal equilibrium in any practical situation (Bruineberg & Rietveld, 2014). This optimal equilibrium can reveal perceptual objects at a maximum of visibility and solicit actions from our bodies to reach such a maximum (Merleau-Ponty, 2012, pp. 315–316).

According to the skilled intentionality framework, the notion of the optimal grip designates the set of conditions that are deemed best for perceptually disclosing the affordances of an object or exploiting these affordances in acting appropriately toward it. Since our actions, as socially enculturated beings, are determined by public standards and not merely by individual goals, our bodies are sensitive to sociocultural norms, that is, to situated normativity. While the constitution of this optimal grip depends on a subject-independent normative framework, the body can become sensitive to the situated normativity of a practice and respond adequately to its solicitations. The body learns how to attune to the demands of a practice. This process of bodily attunement to normative ecological frameworks is skilled intentionality (Bruineberg & Rietveld, 2014).

Skilled intentionality has important similarities to sense-making (Kirchhoff & Froese, 2017) but also significant differences. While skilled intentionality describes a phenomenon of *norm attunement*, sense-making describes one of *norm enactment* (Sepúlveda-Pedro, 2020). In the first case, the body attunes its physical structure and behavior to pre-given and subject-independent normative frameworks. In the second case, the norms are enacted individually or jointly with other agents in the historical development of the agent–environment system (see Sect. 6.3).

The naturalization of skilled intentionality follows Friston's account of the free-energy principle (see Sect. 2.2.2). This principle offers a statistical and dynamical model for understanding how the brain–body–environment system organizes itself to reduce uncertainty or free energy. Uncertainty causes an organizational disequilibrium in the brain–body–environment system that is affectively felt by cognitive subjects as a bodily tension that needs to be reduced (Bruineberg & Rietveld, 2014). The organizational composition of the body and the brain allows subjects to modulate their coupling with the environment to reduce uncertainty. This is possible thanks to a process called active inference.

Active inference produces changes in the system that reorganizes the brain and the body, for example, transforming the subjects' bodily skills and perceptual field (Kirchhoff et al., 2018). Active inference also produces changes in the organization of the environment through motor action. Therefore, the tendency to reduce uncertainty, from a dynamical and statistical point of view, can explain the lived affectivity of the body to reach the optimal grip described by Merleau-Ponty. The affective tendency to the optimal grip is the tendency to reach an equilibrium between the internal states of the brain–body–environment system (i.e., the current self-organization of the system) and the external ones (dynamical changes in the landscape of affordances) (Rietveld et al., 2018, p. 61).

6.2.3 *Norm Attunement*

Although situated normativity can account for the gathering force of place in behavior settings, its current ecological analysis possesses at least two problematic aspects. First, situated normativity portrays a set of norms proper to sociocultural practices, but it does not consider the role that participants can play in the constitution of these norms. This role is not only necessary for explaining the origin of social norms; it can also explain how the body manages to cope with the contingencies of situational contexts that constantly change in bodily practices. Second, situated normativity excludes natural places from community-relative and contextual-relative constraints that characterize sociocultural places, suggesting that general ecological laws rule the relation between an isolated agent and the environment. This picture implies an abstraction that falsely presupposes that living organisms can exist in isolation and that norms of behavior are unaffected by the contingencies of the material contexts that organisms inhabit.

Ecological normative frameworks do not emerge from external processes to the interactional field of the agent–environment system. Whether the interactions of the agent–environment system are between an agent and its physical surroundings or include multiple agents, these ecological frameworks originate principally in the interactions at the system's core. From a third-person perspective, the main features of interactional norms are constituted previously to the participation of agents in practices. However, this fact does not mean that agents blindly follow traditional norms or that they merely get attuned to them. The very nature of bodily practices demands from participants creativity, spontaneity, and managing contingencies in the environment. This condition forces agents to transform already established normative frameworks constantly.

The transformation of norms might seem unnoticeable, but every agent must cope with their individual circumstances and adapt their behavior and bodily organization according to the new situations. Therefore, situated normativity necessarily involves certain stability revealed in behavior patterns observable across the community. However, this normativity also entails lability, allowing for introducing new members to the community while partially transforming the already established norms.

Chapter 5 (Sect. 5.2) described how jazz musicians manage the contingencies of improvisation by reenacting habitual patterns of music into a new order that incorporates these contingencies, thereby affording the possibility of new regular patterns. These patterns can institute new personal and shared styles that change the established normativity of playing jazz. We can see the same creative process in many other artistic styles like painting, literature, sculpture, and architecture. Improvisation and creativity are, however, not exclusive to artistic practices; they are required in all sociocultural practices.

The anthropologist Tim Ingold (2010, 2011) claims that many, if not all, social practices consist of constantly adapting our actions to the unceasing flow of materials and (normative) forces the becoming of the world compels. For Ingold, the paradigmatic example of bodily practices is textile weaving, whereby weavers use available materials to design a unique path of becoming, one that embodies the context of the weavers.

However, it is essential to notice that the dynamic development of norms is not restricted to the sociocultural aspect of bodily practices. Chapter 5 described how goldsmiths need to manage the material's contingencies to give the proper form to jewelry (Sect. 5.2). Indeed, the norms that dictate what counts as good jewelry involve a sociocultural normativity. However, the material's resistance to the habitual sensorimotor skills demands adjustments of the body, independently of the social norm. Animals also manage contingencies in their environment, and in their developmental paths, they learn to adapt their bodies and behavior, thus implementing their vital norms (Oyama, 2000). Bodily practices, sociocultural or not, are therefore highly dynamic interaction processes. For this reason, the situatedness of actions needs to be seen as a highly dynamic field that the analysis offered by the skilled intentionality framework misses.

As we have seen, the skilled intentionality framework recognizes the dynamism of the causal processes that alter the environment's materiality and, consequently, the behavior and perception of agents. The Bayesian models of the free-energy principle explain the constant adjustment of the statistical space caused by these dynamic causal processes (Bruineberg et al., 2018). However, the concept of skilled intentionality does not explain how the interactional norms we observe in these practices originated. It does not explain either why some norms and not others were created if the materiality of the environment allows more than one sociomaterial configuration. Even worse, the notion of skilled intentionality does not explain the logic of the self-transformation of interactional norms. At best, the skilled intentionality framework appeals to the capacity of reflective human consciousness to induce changes in the current field of situated normativity (see, e.g., Rietveld et al., 2017).

On the contrary, the dynamical account of sense-making as a process of norm enactment concerns the norm development and norm envelopment of bodily practices. This enactive shift situates the normativity of agent–environment interactions in a highly dynamical (and conflictive) field. However, it is one with enough stability to sustain regular processes that gives habits, situations, and places their particular style.

As I argued in Chap. 5, the dynamic self-transformation of normative frameworks is not only motivated by the materiality of the environment but also involves the tension between different levels of normativity at play in one single action. The enactive theory of sensorimotor habits, for instance, describes how multiple sensorimotor skills self-organize to coordinate the body's action. However, sensorimotor normativity may become partially independent of biological normativity. This phenomenon produces a tension between the two normative domains, to the point of becoming opposite tendencies, as in bad habits and addictions (Ramírez-Vizcaya & Froese, 2019).

Within the networks of sensorimotor habits and schemes, there may also be tensions that produce a discoordination of bodily actions. It happens, for instance, when we learn a new bodily skill and our habitual repertoire of movements is challenged. Learning to swim, for example, involves, for human adults, a new way to equilibrate their bodies in the water. However, the new bodily equilibrium diverges from the more usual bodily equilibrium when they stand on the ground. In order to learn to swim, a human adult needs to modify her habitual tendency to move her body; only then can she begin to float and progressively acquire new bodily coordination that allows her to dive in water (Cappuccio & Ilundáin-Agurruza, 2020).

Situated normativity is a valuable concept for an account of enactive place. It provides accurate descriptions of the normativity that guides the coordination of agents in bodily social practices and the connection of practical norms to the material locality of places. Nonetheless, the current definitions of situated normativity are not appropriate for the theoretical view of the enactive approach. From their ecological view, the supporters of the skilled intentionality framework presuppose the existence of normative frameworks that are independent of the active engagement and participation of agents. This view is contrary to evidence we have of the dynamic transformations of norms and the material structures of places that occur due to the active participation of agents. We need an account of situated normativity that acknowledges the dynamic processes happening in the interior of places and practices. We need an account of situated normativity in terms of norm enactment (sense-making) rather than norm attunement (skilled intentionality). Fortunately, recent work on the enactive approach to social cognition can help us build an enactive account of situated normativity.

6.3 Place-Norms as Enactive Situated Normativity

In Chap. 5 (Sect. 5.1.2), I described participatory sense-making as the type of sense-making that characterizes social coordination between two or more living agents (De Jaegher & Di Paolo, 2007). The supporters of the enactive approach have moved one step forward from the original model of participatory sense-making to describe the emergence of a new form of agency besides the biological, sensorimotor, and intercorporeal agencies. This new model embodies sociocultural normativity, a fundamental part of linguistic agency (Di Paolo et al., 2018). It describes the emergence of expressive and linguistic utterances that work as tools for regulating and co-regulating bodily actions and interactions between agents. These utterances appear due to dialectical tensions at different stages of metastable processes of interaction and co-regulation between participants of intercorporeal practices. Section 6.3.1 summarizes this model to clarify the dynamic constitution of different levels of social normativity that characterize sociocultural practices from the standpoint of the enactive approach.

This enactive account of social norms is helpful to recognize that social normativity is enactively constituted and not ecologically pre-given, as the descriptions of the skilled intentionality framework suggest. Nonetheless, from the arguments I have been positing in this book, it follows that enactive social normativity is also situated (i.e., they are place-norms). Therefore, the constitution and self-transformation of social normativity (i.e., norm development) also involve the norms enacted in the past of the body–world entanglement (i.e., the habitual body and the active environment), which is sedimented in the spatial localities of agents (Sect. 6.3.2).

6.3.1 The Emergence of Linguistic Bodies

The original model of participatory sense-making already exhibits a permanent tension between two levels of normativity, one relevant to the individual and the other to the social or interactive. In what follows, I describe the model the supporters of the enactive approach have built to understand the emergence of different levels of social normativity that go from simple bodily interactions to linguistic exchanges. All the following descriptions in this subsection are based on the most recent book of the enactive approach: *Linguistic Bodies* (Di Paolo et al., 2018).

In the updated model, this tension remains constant through different stages of conflict (*dissonance*) and harmonization (*synergy*) between the two levels of normativity. Initially, these tensions force individuals to adjust their sensorimotor norms (sensorimotor regulation), but the adjustment must eventually be carried out jointly (sensorimotor co-regulation), creating genuine *social acts*. The sensorimotor co-regulation of social interactions eventually produces social acts that help agents to make the coregulatory acts more efficient. This is a process of meta-coregulation that will be present across the following stages of the model.

The meta-coregulation of social acts implies a new sort of social interaction where a new tension between two levels of normativity is at play. On the one hand, there is the normativity of the practice intended initially by the participants. However, there is a second normativity that involves using partial acts to co-regulate other partial acts. As with any other social practice, coordination of social acts implies the sedimentation of habitual patterns of interaction and creative improvisation with new patterns necessary for the adaptation to the contingencies of all practices. Therefore, there is a *normativity of social interactions* at play in social practices and a *normativity of social acts* that govern how one participant can use patterns of behavior to influence the behavior of other agents.

The regulatory acts that emerge in concrete interactions are first used locally. Some of these acts can reinforce the interactional links between a closed group with a reduced number of members (*local pragmatics*). In contrast, other acts are strongly normative and can regulate acts in a community with a broader number of members. These acts can also be portable among different communities (*portable acts*). As the number of portable acts increases in a community, a mutual reinforcement of social acts establishes a network with closure that comprises *a shared world* for the community members. The recurrent use of portable acts produces what the enactive approach calls *strong normativity*. This normativity creates an asymmetry in the roles of different members in a social group. Some members are active regulators, while others are passively regulated. The roles of *regulator* and *regulated* are constantly interchangeable, though only if the interaction remains social (an interaction between autonomous agents). This interchange of the regulator and regulated roles creates a dialogic dynamic in social interactions. Recognizing the roles of both active and passive regulators allows for identifying others as agents. The member taking the role of the active regulator in a dialogic dynamic makes use of *utterances* (gestural and vocal) to lead and regulate the interaction. These utterances are intrinsically social since their use depends on the relational complementarity between the producer and the audience.

The efficiency of social acts of co-regulation and meta-coregulation leads agents to recognize each other as agents mutually. This phenomenon becomes evident in the emergence of a dialogic interaction where the roles of an active regulator and a passive regulated member are interchangeable. At this dialogical level, agents use utterances to regulate social interactions. There is a progressive construction of dialogical networks of utterances shared by a community in particular contexts of bodily actions. Di Paolo and colleagues call these networks *participation genres*.

Participation genres resonate with the notion of microworlds at the sensorimotor level of autonomy. However, in participation genres, the normative structures of interaction involve the networks of dynamic sensorimotor processes and a network of utterances. These networks are constantly regulated by processes of ***mutual interpretation*** between multiple participants. However, these regulations can take the form of self-interpretation when producers of utterances become aware of the impairment between the pragmatic and expressive aspects of her utterances, such as when the utterance does not produce the expected responses in others. This moment is crucial because the model incorporates a new level of reflective and dialogical dynamics to the intersubjective skills of an agent. Successful utterances become regular patterns of dialogical practices for interactions with other agents and themselves. The norms, co-enacted with others and embodied in networks of utterances (participation genres), now play a more explicit role as a tool for self-regulation. Agents incorporate utterances as regulatory tools for their expressive and pragmatic goals. As tools for self-reflection or *self-control*, these new dialogical networks afford the possibility of questioning and making explicit the already existing normativities that agents can now reshape and move forward dialogically.

6.3.2 From Social to Enactive Situated Normativity

Many supporters of ecological approaches criticize the enactive approach for being incapable of explaining social normativity. The critique targets their account of vital, sensorimotor, and intercorporeal forms of sense-making, which would exclusively refer to the normative domain of individuals (Heras-Escribano et al., 2013, 2015). In response, scholars of the enactive approach have developed a theoretical sketch of the emergence of social norms arising from tensions inherent to the social interactions of autonomous agents. In this model, it is possible to recognize how individual agents progressively acquire new regulatory processes that emerge from social interactions. While it is only at the final stages that social and public norms more explicitly guide agents' actions, since the early stages of participatory sense-making, norms jointly enacted by more than one individual constrain the embodiment of individual agents.

The model of norm attunement of the skilled intentionality framework assumes that ecological normative frameworks are pre-given or constituted by processes external to the bodily interactions of practice participants. The enactive model, by contrast, explains how sociocultural frameworks are constituted by the participation of agents in both local and wider communities of individuals. The participation of agents at the different levels of constitution of shared processes of regulation and co-regulation implies that an internal dynamic tension is always present in sociocultural practices. These tensions can explain the self-transformation of the normative frameworks, not only because of the changes in the materiality of sociocultural environments but also due to the dynamical interactions happening "inside" of the sociocultural practices.

Recall the above example of jazz musicians co-enacting musical patterns that institute new personal, group, and genre styles (Sects. 5.2.2 & 5.2.3). The temporal unfolding of collective jazz improvisation is affected by the process occurring in the interior of the interactional domain between musicians and by the spatial and temporal surroundings of this practice, for example, the public, the hall, musicians' moods, historical events, etc. The interaction determines the sense of the practice. The self-transformation of the practice at different temporal scales is reached thanks to these internal and external dynamics. There is a metastability necessary for the communion of agents in a shared practice and involving different levels of normativity, from the biological affectivity of agents to sensorimotor and sociocultural norms.

A careful and detailed analysis between this linguistic model and practices like jazz improvisation remains to be carried out. However, the linguistic theory of the enactive approach seems to miss again the environment, which appears to be only in the background of the interactional space of agents and not entangled within the intercorporeal practices. The entanglement of communal interactions and the materiality that flows both out from and inside the interactional space of musicians requires a complex and detailed explanation that exceeds the scope of this work. There are, however, already clues as to the potential of the enactive approach for explaining the role of music creation, learning, and appreciation (Schiavio & Cummins, 2015; Schiavio & De Jaegher, 2017; Van der Schyff et al., 2018).

These are also essential characteristics of the dynamic constitution of place that I have previously shown in the ecological web of *chinampas* (Sect. 6.1.2). Interaction of agents and nature sediments material and normative structures that constrain and motivate the self-transformation of chinampas, for example, from agriculture fields to gardens and foundations of imperial cities. The temporal scale and, consequently, the dynamicity of *chinampas* are different from the jazz musical space. The complexity of *chinampas* also increases since the interactions with nonhuman agents are also involved in the constitution of these places. However, we can see general affinities in their constitution that help us define places due to dynamic and enactive processes.

Both *chinampas* and collective jazz improvisation are, however, sociocultural places. According to my arguments, the enactive approach can explain situated normativity at the sociocultural level. However, it remains unanswered if there is a situated normativity that applies to pre-cultural bodily practices or a sort of situated normativity that can characterize natural places. The final section will address this subject.

6.4 From Social to Natural Places

Although I have described enactive places based fundamentally on sociocultural places and behavior settings, I do not think situated normativity is a phenomenon exclusive to human culture. It is hard to think of any biological agent constituting its realm of interactions with the environment without the intervention of interactional processes with other agents. There are many processes of social interaction occurring in nature that are intercorporeal and might involve activities of participatory sense-making. Most of these processes do not reach the complex levels of co-regulation and meta-co-regulation of human societies. Still, they are processes that constitute a shared or communal normative field for multiple organisms and species. Therefore, it seems reasonable to imagine a socially situated normativity exists in nature as well.

In this last section, I first survey some ideas of Merleau-Ponty's philosophy of nature. He thought of nature as a symphony where all living and nonliving beings are interconnected. This idea is quite similar to the view we can build from an enactive perspective once we recognize the emplacement of the living bodies (Sect. 6.4.1). Even though we can hypothesize the existence of place and a situated normativity in nature as well as in culture, we need to acknowledge the significant differences between the enacted normative domains of humans and those of the rest of the living organisms (Sect. 6.4.2).

6.4.1 Intersubjectivity, Intercorporeality, and Interanimality

Ecological approaches recognize the situated normativity involved in sociocultural places. Nevertheless, these approaches usually deny that similar forms of normativity can explain the constitution of the local domains of nonhuman animals, which they refer to as ecological niches. For ecological approaches, these niches are constituted by genomic and extra-genomic inheritance processes, where agents' individual development still seems unimportant (e.g., Heras-Escribano & de Pinedo, 2018). This vision contrasts with the conception of life supported by the enactive approach and the definition of cognition as sense-making (see Sect. 3.4). I have argued that the process of norm enactment and sense-making offers a better account of norms of action and perception than the one of norm attunement and skilled intentionality. It remains to be seen if the process of norm enactment is also a better or at least a viable approach for defining the enactment of a situated normativity in natural places.

The claim that situated normativity exists in natural places can challenge the belief that mechanical processes constitute living agents and their ecological niches. From a retrospective look and at an evolutionary timescale, mechanistic explanations seem plausible and sufficient for explaining the adaptation of organisms to their niches. Still, we can also hypothesize that such adaptation could involve the creative improvisation of the ancestors of a species and processes of constitution for joint action between multiple agents that coevolve as phylogenetic groups along with the resilient unities of place. The proper treatment of this subject requires an independent analysis that confronts traditional conceptions of evolutionary biology and the enactive perspective, based on a vision that focuses on development as the key to understanding life and evolution. In this section, however, I want to show that we have reasons to think that normativity in pre-cultural forms of life and natural places also determines their interactions.

Cummins and De Jesus (2016) have criticized the enactive approach for defining life based on agents' autonomy alone. Heras-Escribano et al. (2015) have also questioned whether autonomy can constitute the norms of animal behavior. According to my redefinition of sense-making as norm development, it should be clear that I am also against the idea that the autonomy of life *alone* can constitute the norms of interaction in an agent–environment system. In contrast to the traditional criticisms of the enactive approach, I think that autonomy and agency are constitutive aspects of life, but we also need to offer an account of the emplacement of agents.

The body–world entanglement transcends spatially and temporally the history of any agent–environment system because the enactment of norms depends on already established normative frameworks or levels. Archeologically, we can unearth the past and question what transcends the constitution of an autonomous system and its history of concrete entanglement with the world. We can start by saying that there were other living agents because life, like sense, cannot come from anything. Life implies various reproductive processes or *a transgenerational self-transformation of the body–world entanglement* that makes the birth of new agent–environment systems possible.

There is, therefore, a natural history, one that cannot produce a narrative on its own until beings like us start to make use of symbolic language. There is, however, a temporal connection at the level of life and nature that we can study with the help of contemporary theories of life and evolution (Thompson, 2007) and probably a new philosophy of nature (Gallagher, 2017).

Human history is characterized by social processes that imply significant interactions among different social groups: economic, political, and behavioral. It is common to believe that natural history involves different sorts of temporal processes. These processes concern neutral physical compositions, complex systems like genomes, or cellular and multicellular organisms. From the perspective of the enactive approach, I think it is more proper to think that natural history resembles human history. Nature is also inhabited by individuals that participate in societies that constitute wild places and whose participation actively transforms the course of their natural groups, first locally and eventually globally.

Participatory sense-making is the sort of description that can help us to understand the emergence of a situated normativity at the level of nature that is also socially determined. That is, socially constrained behavior might not be a feature exclusive to human beings.

Ants interact socially and participate in collective actions to maintain the stability of an emergent social system: the anthill (Gordon, 1996; Stapleton & Froese, 2015). There is a distribution of labor in the hill where ants interact to sustain what is sometimes called a superorganism (Canciani et al., 2019). The definition of a superorganism is controversial, but the emplacement of ants' society should not be. The life of individual ants depends on the self-sustaining processes of the collective system. The constitution of the collective system relies on the action of the individual members of the collectivity. The collective ant system is more than its parts since the self-sustaining processes of the system do not depend on each individual in a specific manner (Detrain & Deneubourg, 2006). However, the behavior of individual ants is constrained by the systemic equilibrium of the hill (cf. Anderson & McShea, 2001). We still need more evidence, but we might be talking about processes of participatory sense-making at work in the interactions of ants working in a society that an ant colony represents.

Most importantly, the hill as the place of social interaction is central to regulating the collective systems' behavior. Suppose the hill suffers damage due to natural causes or human interventions. In that case, ants will behave differently and start self-organizing to reconstruct their society and the place where this society lives. Ants and their habitat are also entangled by a normativity that involves the self-organizing processes of their societies. These processes very likely involve a participatory sense-making of the colony individuals.

An anthill and a beehive are just paradigmatic examples of ecological niches that involve niche construction processes or the active transformation of the material surroundings by living agents, making the entanglement of organisms and the environment more evident. However, sociality and niche construction are common processes of nature (Odling-Smee et al., 2003; Turner, 2009). Ecological niches are, in the end, natural places.

Natural places involve the interrelational web of living and nonliving things in a locality that maintains normative co-regulation processes. These processes do not only involve interaction between members of the same species, like in the case of ants and bees but across different species that maintain a systemic equilibrium in a delimited spatial area or location (Ingold, 2011). This area is a place, and like social places, it does not have clear boundaries and depends on processes that transcend the locality of the ecological niches. There are, however, interrelated processes that help for the self-sustaining processes of equilibrium of the niche (cf. Scheffer & van Nes, 2006), so we can talk about the same place, like in the case of *chinampas*. The complexity of the constitutional process of this ecological niche is enormous to be more than intuited at this point. However, we can observe the interdependence of species in the interactions that the different organism maintains in the *chinampa's* niche.

From all that has been said thus far, it should be clear that natural places are not meaningless realms that acquire form or meaning thanks to the organizational interiority of the living body. On the contrary, the enactment of the (minimal) self of the lived body (interiority) and the *Umwelt* (exteriority) occurs as a result of dynamic and dialectic processes that occur within and outside the boundaries of the operational closure of the living body. Therefore, it is not only the autonomous organization of a living being that cognition requires but also the emergent constraints that shape the unfolding of this autonomous organization—and this involves the environment as well.

If these interactional emergent processes are effectively a requirement for life and cognition, we must reject the idea that an *Umwelt* is a mere bubble-like domain, the most common metaphor used to describe *Umwelten* (Sect. 4.1.1). Instead, *Umwelten* are more accurately represented by the metaphor of rings that overlap each other (Buchanan, 2008; Merleau-Ponty, 1995). The idea is that the life of organisms is always interrelated with many other organisms of the same and different species. These organisms and their interactions shape and reshape the place that all of them inhabit. At this point, we can use an analogy that in social cognition is described as an *intercorporeality*, and one named by Merleau-Ponty, at the level of life and nature, as *interanimality* (Merleau-Ponty, 1995, p. 374). Interanimality expresses how animals are entangled in a complex system of relations constitutive of their own being (Toadvine, 2009).

There is another helpful metaphor to depict these *Umwelten*. Merleau-Ponty (1942), following Uexküll, described living beings as melodies or as temporally extended wholes that sing themselves. This metaphor fits perfectly with the enactive approach descriptions of living organisms (Thompson, 2007). However, we must aggregate that these melodies are played over an underlying harmony that interrelates the melodic organisms,[3] producing a symphony with no director nor any *ready-made* score to follow. Instead, like jazz improvisation, these melodies unfold under the current dialogue of music, creating new structures or scaffoldings that disclose new paths for the continuity of the musical intercourse.

The construction of the musical interwoven of nature is not reducible to the developmental history of organisms. As I said, it necessarily involves a phylogenesis and an embryogenesis that constitute very likely the first kind of social relationship between organisms, a transgenerational relation. We can, for now, speculate that the link between a progenitor-organism and a newborn organism can be crucial to shape (and reshape) the bodily constitution of both organisms. Such a relation could also trigger constitutive changes in the sense-making of both organisms. That is, reproduction may imply some basic form of participatory sense-making. Cummins and De Jesus have a point when they insist that sociality is pervasive in life, from bacteria to insect colonies, from flocks of birds to primates' tribes, and from the human microbiome to the constitution of multicellular organisms (Dupré, 2012). The most important conclusion is that forms of life depend on their mutual relations, and it is hard to see how the picture of a solitary organism is an accurate depiction of life.

The example of chinampas is, for this reason, illustrative of an enactivist theory of place. In *chinampas*, as described in Sect. 6.1.2, there is a deep entanglement in the life of human and nonhuman agents. The mold in the bottom of the lake, necessary for the agriculture, depends on the water provided by the *Cerro de la Estrella*, the microbes of the soil, and the activity of the almost extinguished axolotls. The *ahuejotes* support the borders of *chinampas*, but farmers need to plant these trees in a specific location to maintain the stability of the bed of earth. Bugs and birds threaten the harvest of crops. However, since they give balance to the organismic population of the lake, the farmers avoid the use of pesticides and use other useful techniques to control their activity. The *chinamperos* constantly change their life to resist the damages caused by industrial capitalism and fight to preserve the increasingly fragile ecological balance in the lake of Xochimilco. Thus, *chinampas* compose a melodic symphony; one full of harmony, rhythms, breaks, and tensions that evolves as a resilient systemic structure and gives shape to the landscape we currently observe in Xochimilco.[4]

A complete account of sense-making as place-norms needs to keep inquiring further about the processes of birth, genesis, and self-transformation of individuals, groups, species, niches, and habitats, at the different timescales of life and nature. Despite its complexity, I am convinced that we must study living and cognitive phenomena within the historical entanglement of places and their living bodies.

6.4.2 A Jointly Enacted Objectivity

A situated normativity in nature does not impede us from recognizing that significant differences exist between the world lived and dwelled by human agents and the local regions inhabited by nonhuman organisms. Our bodies and brains' complexity and the surplus of culture and language radically transform our normative domain of interaction with the world. From the point of view of multiple enactive and ecological approaches (Di Paolo et al., 2018; Gallagher, 2017; Hutto & Myin, 2017; Kiverstein & Rietveld, 2018), social interactions, along with the development of the human brain, are the main sources of the sophisticated forms of cognition humans possess. This last section briefly surveys some differences between the situated normativity of nature and the one that is more proper of human culture.

The capacity to constitute an objective world separates human cognition and experience from that of other living agents. Merleau-Ponty already acknowledged this fundamental difference in the *Structure of Behaviour* (Merleau-Ponty, 1942). Perceptually, for instance, we can detach objects from their immediate practical context and from an egocentric orientation of the world (i.e., how the environment is situated in reference to our own body), whereas nonhuman animals cannot. Despite the proximity of their intelligence to ours, chimpanzees cannot solve problems that require an appreciation of the standpoint of the object, and not simply from the egocentric perception of the chimpanzee (Kohler, 1999).

For the enactive approach, intersubjective interactions in humans also develop more abstract perceptual attitudes. Intersubjective interactions between caregivers and children enable the latter to understand the perspective of an object from a different point of view than their own (Di Paolo et al., 2017). Here, intersubjectivity decentralizes the contents of perception away from the ego pole, making it possible to see an object independently of our immediate pragmatic concerns.

This kind of abstract object perception also involves complex forms of social processes of co-regulation (Di Paolo, 2016). For the enactive approach, the perception of an object is always (at least implicitly) shaped by actual or possible intersubjective interactions. Hence, the meaning or the normative value of the object is not simply determined by the history of the agent–environment coupling but also by the history of coordination and conflicts (through corporeal interactions) with others. The meaningful environment or the interactive space of two or more agents is thus signified according to the mutual co-regulation of multiple agents and by the social autonomous organization that emerges from their interaction (Di Paolo et al., 2018).

It appears that most animals are also incapable of transposing from the perceptual level of an object to the imaginative level of cognition. This phenomenon happens, for instance, when human children start using objects as if they were another sort of object. In order to conceive of imagination, from an enactive perspective close to that of the enactive approach, Gallagher (2017) proposes to follow Gilbert Ryle's account of imagination as simulation and pretense, albeit in the form of an expansion of the affordances available in the environment.

For Gallagher, human cognitive activities like imagination expand the realm of affordances from concrete situations, in this case, to imaginary situations (Gallagher, 2017, p. 194). This idea becomes more precise when Gallagher uses the example of a child playing with a banana as a substitute for a phone (Sainsbury, 2009). Gallagher sees this action as involving the child finding similar affordances in the banana and the phone, thereby allowing the child to be physically engaged in a similar manner with both objects, applying thus a metaphoric switch from a real situation to an imaginary one. We should notice that, for Gallagher, this metaphoric switch occurs in terms of the whole brain–body–environment system since the substitution occurs in the concrete perception–action organization (a Gestalt) and not just in the interconnection of brain states (Gallagher, 2017, p. 195).

The importance of Gallagher's account of imagination lies in his overall redefinition of imagination as embodied action and the manipulation of environmental resources, including ideal objects such as concepts. For Gallagher, concepts are objects that offer us new affordances. They open new possibilities of acting with and manipulating the environment. Concepts, however, are not mere abstractions taking place in the head; instead, they are embodied in language, images, and symbols (Gallagher, 2017, p. 196). Indeed, the active manipulation of the environment helps us develop our most sophisticated cognitive skills, such as mathematical thinking. For instance, using our bodies and tools is helpful and necessary for developing mathematical reasoning (Gallagher, 2017, pp. 210–212). Using our fingers or an abacus to count are simple illustrations of this idea.

In short, these arguments suggest that humans can detach specific aspects of the environment from their proximate spatiotemporal locations, relocating them in abstract situational contexts. Symbolic language and culture undoubtedly enable many other ways of reconfiguring the horizons of our perception to create new normative levels of cognition that nonetheless remain entangled with the lower levels, keeping attached ourselves to the world of nature.

Notes

1. Although less visible and thick, all sort of places should possess characteristics similar to those I will describe in the *Chinampas* example. Our contemporary cities, buildings, and houses, for instance, also have a temporal unfolding that entangles with the history of interactions of human and non-human populations (Norman, 2013).
2. The name *chinampas*, means in Nahuatl (Mexicas' language) surrounded by canes and hedges (*chinánitl*) and above the surface of water (*pan*) (Frederick, 2007, p. 110).
3. We usually associate harmony with the smooth coordination of multiple components in a systemic whole. However, musical harmony underlay many synchronic and diachronic tensions between its components (e.g., chords and progressions). These musical tensions resonate with the conflicts described by the enactive approach in the processes of coordination, breakdowns, and recoveries of autonomous systems and social interactions (Di Paolo et al., 2018).
4. A place is only meaningful to the observer that can comprehend its existential unity. The eye of the bourgeois, for instance, can look differently and see nothing in *chinampas* but opportunities or impediments to make his profits. The politician looking indifferently to the population's way of life cannot see the richness of the lacustrine environment of Xochimilco; the fertile land, the colorful festivals, the ancient rituals, the history, and the significance of the many forms of life that inhabit this place. Hence, place is a unity, but we cannot see this unity if we do not comprehend, make sense, and take care of place as a precinct of life and culture.

References

Anderson, C., & McShea, D. W. (2001). Individual versus social complexity, with particular reference to ant colonies. *Biological Reviews, 76*(2), 211–237.

Aveni, A. F., Calnek, E. E., & Hartung, H. (1988). Myth, environment, and the orientation of the Templo mayor of Tenochtitlan. *American Antiquity, 53*(2), 287–309.

Barker, R. G. (1968). *Ecological psychology: Concepts and methods for studying the environment of human behavior.* Stanford University Press.

Barker, G. (2017). From stability to norm transformation: Lessons about resilience, for development, from ecology. *Phenomenology and the Cognitive Sciences, 16*(4), 571–584.

Bruineberg, J., Chemero, A., & Rietveld, E. (2018). General ecological information supports engagement with affordances for 'higher' cognition. *Synthese, 196*(12), 5231–5251. https://doi.org/10.1007/s11229-018-1716-9

Bruineberg, J., & Rietveld, E. (2014). Self-organization, free energy minimization, and optimal grip on a field of affordances. *Frontiers in Human Neuroscience, 8*, 599. https://doi.org/10.3389/fnhum.2014.00599

Buchanan, B. (2008). *Onto-ethologies: The animal environments of Uexküll, Heidegger, Merleau-Ponty, and Deleuze*. SUNY Press.

Calnek, E. E. (1972). Settlement pattern and chinampa agriculture at Tenochtitlan. *American Antiquity, 37*(1), 104–115.

Canciani, M., Arnellos, A., & Moreno, A. (2019). Revising the superorganism: An organizational approach to complex eusociality. *Frontiers in Psychology, 10*, 2653.

Cappuccio, M. L., & Ilundáin-Agurruza, J. (2020). Swim or sink: Habit and Skillful control in sport performance. In F. Caruana & I. Testa (Eds.), *Habits: Pragmatist approaches from cognitive science, neuroscience, and social theory* (pp. 137–162). Cambridge University Press.

Casey, E. S. (1996). How to get from space to place in a fairly short stretch of time: Phenomenological prolegomena. In S. Feld & K. Basso (Eds.), *Senses of place* (pp. 13–52). School of American Reserch.

Casey, E. S. (1998). *The fate of place: A philosophical history*. University of California Press.

Charli-Joseph, L., Siqueiros-Garcia, J. M., Eakin, H., Manuel-Navarrete, D., & Shelton, R. (2018). Promoting agency for social-ecological transformation. *Ecology and Society, 23*(2), 46. https://doi.org/10.5751/ES-10214-230246

Chemero, A. (2009). *Radical embodied cognitive science*. MIT Press.

Cox, A. M., Martins, J. T., & González, G. R. (2020). Reassessing the LIS approach to traditional knowledge: Learning from Xochimilco, Mexico city. *Journal of Documentation., ahead-of-print* (ahead-of-print), *76*, 981. https://doi.org/10.1108/jd-10-2019-0195

Crossley, P. L. (2004). Just beyond the eye: Floating gardens in Aztec Mexico. *Historical Geography, 32*, 111–135.

Cummins, F., & De Jesus, P. (2016). The loneliness of the enactive cell: Towards a bio-enactive framework. *Adaptive Behavior, 24*(3), 149–159.

Dawkins, R. (2006). *The selfish gene*. Oxford University Press.

De Jaegher, H., & Di Paolo, E. A. (2007). Participatory sense-making. *Phenomenology and the Cognitive Sciences, 6*(4), 485–507. https://doi.org/10.1007/s11097-007-9076-9

de Ruiz, S. L. (1973). *Desarrollo urbano de México-Tenochtitlán según las fuentes históricas 3*. Instituto Nacional de Antropología e Historia.

Detrain, C., & Deneubourg, J.-L. (2006). Self-organized structures in a superorganism: Do ants "behave" like molecules? *Physics of Life Reviews, 3*(3), 162–187.

Di Paolo, E. A. (2016). Participatory object perception. *Journal of Consciousness Studies, 23*(5–6), 228–258.

Di Paolo, E. A., Buhrmann, T., & Barandiaran, X. E. (2017). *Sensorimotor life: An enactive proposal*. Oxford University Press.

Di Paolo, E. A., Cuffari, E., & De Jaegher, H. (2018). *Linguistic bodies: The continuity between life and language*. MIT Press.

Dupré, J. (2012). *Processes of life: Essays in the philosophy of biology*. Oxford University Press.

Eakin, H., Shelton, R. E., Siqueiros-Garcia, J. M., Charli-Joseph, L., & Manuel-Navarrete, D. (2019). Loss and social-ecological transformation. *Ecology and Society, 24*(3), 1.

Frederick, C. D. (2007). Chinampa cultivation in the basin of Mexico. In T. L. Thurston & C. T. Fisher (Eds.), *Seeking a richer harvest* (pp. 107–124). Springer.

Friston, K. J. (2010). The free-energy principle: A unified brain theory? *Nature Reviews Neuroscience, 11*(2), 127–138.

Gallagher, S. (2017). *Enactivist interventions: Rethinking the mind*. Oxford University Press.

Gliessman, S. R., Garcia, R. E., & Amador, M. A. (1981). The ecological basis for the application of traditional agricultural technology in the management of tropical agro-ecosystems. *Agro-Ecosystems, 7*(3), 173–185.

Gordon, D. M. (1996). The organization of work in social insect colonies. *Nature, 380*(6570), 121–124.

Heft, H. (2018). Places: Widening the scope of an ecological approach to perception–action with an emphasis on child development. *Ecological Psychology, 30*(1), 99–123.

Heras-Escribano, M., & de Pinedo, M. (2018). Affordances and landscapes: Overcoming the nature–culture dichotomy through niche construction theory. *Frontiers in Psychology, 8*, 2294.

Heras-Escribano, M., Noble, J., & de Pinedo, M. (2013). *The only wrong cell is the dead one: On the enactive approach to normativity* (pp. 665–670). MIT Press. https://doi.org/10.7551/978-0-262-31709-2-ch095

Heras-Escribano, M., Noble, J., & de Pinedo, M. (2015). Enactivism, action and normativity: A Wittgensteinian analysis. *Adaptive Behavior, 23*(1), 20–33.

Hutto, D. D., & Myin, E. (2017). *Evolving enactivism: Basic minds meet content*. MIT Press.

Ingold, T. (2010). The textility of making. *Cambridge Journal of Economics, 34*(1), 91–102.

Ingold, T. (2011). *Being alive: Essays on movement, knowledge and description*. Taylor & Francis.

Kant, I. (1998). In P. Guyer & A. W. Wood (Eds.), *Critique of pure reason*. Cambridge University Press.

Kirchhoff, M. D., & Froese, T. (2017). Where there is life there is mind: In support of a strong life-mind continuity thesis. *Entropy, 19*(4), 169.

Kirchhoff, M. D., Parr, T., Palacios, E., Friston, K., & Kiverstein, J. (2018). The Markov blankets of life: Autonomy, active inference and the free energy principle. *Journal of the Royal Society Interface, 15*(138), 20170792.

Kiverstein, J. D., & Rietveld, E. (2018). Reconceiving representation-hungry cognition: An ecological-enactive proposal. *Adaptive Behavior, 26*(4), 147–163.
Kohler, W. (1999). *The mentality of apes.* (E. Winter, Trans.). Routledge.
Laland, K. N., Odling-Smee, F. J., & Feldman, M. W. (2000). Niche construction, biological evolution, and cultural change. *Behavioral and Brain Sciences, 23*(1), 131–146.
Lewontin, R. C. (1983). The organism as the subject and object of evolution. *Scientia, 118*, 63–82.
Malpas, J. E. (2004). *Place and experience: A philosophical topography.* Cambridge University Press.
McGann, M. (2014a). Enacting a social ecology: Radically embodied intersubjectivity. *Frontiers in Psychology, 5*, 1321.
McGann, M. (2014b). Situated agency: The normative medium of human action. *Synthesis Philosophica, 58*(2), 217–233.
Merleau-Ponty, M. (1942). *The structure of behaviour.* (A. L. Fisher, Trans.). Duquesne University Press.
Merleau-Ponty, M. (1995). In D. Séglard (Ed.), *La Nature: Notes, Cours du Collège de France.* Editions du Seuil.
Merleau-Ponty, M. (2012). *Phenomenology of perception.* (D. Landes, Trans.). Routledge.
Merlín-Uribe, Y., González-Esquivel, C. E., Contreras-Hernández, A., Zambrano, L., Moreno-Casasola, P., & Astier, M. (2013). Environmental and socioeconomic sustainability of chinampas (raised beds) in Xochimilco, Mexico City. *International Journal of Agricultural Sustainability, 11*(3), 216–233.
Morehart, C. T., & Frederick, C. (2014). The chronology and collapse of pre-Aztec raised field (chinampa) agriculture in the northern basin of Mexico. *Antiquity, 88*(340), 531.
Morris, D. (2004). *The sense of space.* State University of New York Press.
Morris, D. (2018). *Merleau-Ponty's developmental ontology.* Northwestern University Press.
Narchi, N. E., & Cristiani, B. C. (2015). Subtle tyranny: Divergent constructions of nature and the erosion of traditional ecological knowledge in Xochimilco. *Latin American Perspectives, 42*(5), 90–108.
Newton, I. (1999). *The principia: Mathematical principles of natural philosophy.* (B. Cohen & A. Whitman, Trans.). University of California Press.
Norman, D. (2013). *The design of everyday things: Revised and expanded edition.* Basic books.
Odling-Smee, F. J., Laland, K. N., & Feldman, M. W. (2003). *Niche construction: The neglected process in evolution (MPB-37).* Princeton University Press.
Oyama, S. (2000). *The ontogeny of information: Developmental systems and evolution* (2nd ed.). Duke University Press.

Ramírez-Vizcaya, S., & Froese, T. (2019). The enactive approach to habits: New concepts for the cognitive science of bad habits and addiction. *Frontiers in Psychology, 10*, 301.

Rietveld, E. (2008). Situated normativity: The normative aspect of embodied cognition in unreflective action. *Mind, 117*(468), 973–1001.

Rietveld, E., Denys, D., & Van Westen, M. (2018). Ecological-enactive cognition as engaging with a field of relevant affordances. In L. de Bruin, A. Newen, & S. Gallagher (Eds.), *The Oxford handbook of 4E cognition* (p. 41). Oxford University Press.

Rietveld, E., & Kiverstein, J. D. (2014). A rich landscape of affordances. *Ecological Psychology, 26*(4), 325–352.

Rietveld, E., Rietveld, R., & Martens, J. (2017). Trusted strangers: Social affordances for social cohesion. *Phenomenology and the Cognitive Sciences, 18*(1), 299–316.

Sainsbury, R. M. (2009). *Fiction and fictionalism*. Routledge.

Scheffer, M., & van Nes, E. H. (2006). Self-organized similarity, the evolutionary emergence of groups of similar species. *Proceedings of the National Academy of Sciences, 103*(16), 6230–6235.

Schiavio, A., & Cummins, F. (2015). In R. Timmers, N. Dibben, Z. Eitan, R. Granot, T. Metcalfe, A. Schiavio, & V. Williamson (Eds.), *An inter (en) active approach to musical agency and learning* (Vol. 201).

Schiavio, A., & De Jaegher, H. (2017). Participatory sense-making in joint musical practices. In M. Lessaffre, P.-J. Maes, & M. Leman (Eds.), *The Routledge companion to embodied music interaction* (pp. 31–39). Routledge.

Sepúlveda-Pedro, M. A. (2020). Levels and norm-development: A phenomenological approach to enactive-ecological norms of action and perception. *Frontiers in Psychology, 11*, 1666. https://doi.org/10.3389/fpsyg.2020.01666

Stapleton, M., & Froese, T. (2015). Is collective agency a coherent idea? Considerations from the enactive theory of agency. In C. Misselhorn (Ed.), *Collective agency and cooperation in natural and artificial systems* (Vol. 122, pp. 219–236). Springer.

Thompson, E. (2007). *Mind in life: Biology, phenomenology, and the sciences of mind*. Harvard University Press.

Toadvine, T. (2009). *Merleau-Ponty's philosophy of nature*. Northwestern University Press.

Torres-Lima, P., Canabal-Cristiani, B., & Burela-Rueda, G. (1994). Urban sustainable agriculture: The paradox of the chinampa system in Mexico City. *Agriculture and Human Values, 11*(1), 37–46.

Turner, J. S. (2009). *The extended organism: The physiology of animal-built structures*. Harvard University Press.

Van der Schyff, D., Schiavio, A., Walton, A., Velardo, V., & Chemero, A. (2018). Musical creativity and the embodied mind: Exploring the possibilities of 4E cognition and dynamical systems theory. *Music & Science, 1*. https://doi.org/10.1177/2059204318792319

van Dijk, L., & Rietveld, E. (2017). Foregrounding sociomaterial practice in our understanding of affordances: The skilled intentionality framework. *Frontiers in Psychology, 7*, 1969.

Wittgenstein, L. (2007). *Lectures and conversations on aesthetics, psychology and religious belief.* University of California Press.

Wittgenstein, L. (2009). *Philosophical investigations.* (G. E. M. Anscombe, P. M. S. Hacker, & J. Schulte, Trans., P. M. S. Hacker & J. Schulte, Eds.) (4th ed.). Wiley-Blackwell.

CHAPTER 7

Finale: Situating the Enactive Approach

This book began by asking about the role the environment plays in the constitution of processes of life and cognition. The arguments I have made throughout are directed toward the thesis that, from the perspectives of the enactive approach, and from those of phenomenology, we are embodied beings deeply entangled with our natural and cultural surroundings. Our entanglement with the environment not only comprises the causal dependency of our bodies on the natural and material world, but also our historical intertwining with many processes around us, including the embodied existence of multiple living beings.

This view is concurrent with the cosmovision of many autochthonous communities that acknowledge their attachment to the natural and cultural places where they live. It contrasts with the vision of modern capitalist societies that have lost track of their connection to their proximate niches, and which treat the environment as something alien, manipulable, exploitable, and replaceable. Between these two antagonistic views, it is possible to build a new way to relate to the world, one that facilitates working together as a global civilization that must work jointly with nature to overcome the challenges we face. While it is true this book is not directly concerned with the ethical and urgent practical implications of ecological thinking, it should be clear that the philosophical and scientific views of

M. A. Sepúlveda-Pedro, *Enactive Cognition in Place*, New Directions in Philosophy and Cognitive Science,
https://doi.org/10.1007/978-3-031-20282-7_7

the enactive approach, ecological psychology, and finally, of phenomenology invite reflection on the things we need to change to save our species, and our natural world.

More specifically, we've seen that the enactive approach addresses many problems in cognitive science from a non-reductionist perspective. Believing that our minds are based on nonmechanistic principles, the enactive approach redefines our relationship with the world around us. Critically, the enactive approach holds that the environment is not alien to us, because it mirrors the development of our biological, sensorimotor, social, and linguistic bodies. There is an interdependence and codetermination of the agent and the environment that emerges as sense-making acts in processes of life and cognition. Such a description of the agent–environment relationship helps us adopt a wider perspective on the many factors that shape and constitute our cognitive functions. However, sense-making is not a simple description of functional processes; instead, sense-making should be understood as pervasive in our lives. It is there every time we breathe, walk, hear, think, care, love, and hate. Sense-making is ultimately the emergence of the orientation and meaning of our individual and collective everyday actions.

From this enactive stance, the emplacement of living bodies is a crucial dimension for study and analysis. Understanding the deep connection of our bodies to the different layers of the environment is a complex matter, and this book has barely sketched the connections between our bodies to the many dimensions of the environment.

This final chapter looks at some of the different lines of research that lie ahead, while concluding some of the main ideas I've proposed to improve the view of the ecological dimension of cognition from the perspective of the enactive approach. Section 7.1 looks at the implications of thinking about cognition, while taking into account the fundamental circularity that unavoidably characterizes the study of cognition itself. In Sect. 7.2, I summarize the main ideas I have suggested to help shift our conception of sense-making and the active role the environment plays in this alternative view of enactive cognition. In Sect. 7.3, I clarify my proposal to understand sense-making as norm development, and to conceive the environment as an active ecological field. Finally, in Sect. 7.4, I survey the possible applications of the study of place, and the challenges to materializing the study of such a complex field.

7.1 The Fundamental Circularity of Enactive Cognition

Since the 1990s, there has been an ongoing transformation of cognitive science. Much of this transformation concerns the roles we grant to the body and to the environment in our scientific explanations of cognitive phenomena. While this transformation should, in some scholars' eyes, consist of a reform of the already established theories and methodologies of brain-centered cognitive science, others argue that this transformation needs to be more revolutionary, and shift cognitive science into a new scientific paradigm. That latter vision is what guides world-involving cognition. This revolutionary enterprise requires a radical replacement of the main theoretical assumptions, mathematical tools, and observational methodologies in the field.

In the first three chapters of this book, I argued in favor of one particular vision of world-involving cognition, namely, the enactive approach. This approach is the most radical form of enactive cognition because it pushes us to reassess our presuppositions about the definitions and explanations of mind, body, and of the environment in cognitive science. The theory of biological autonomy and the concept of sense-making, both elements making up the theoretical core of the enactive approach, are controversial and provocative for this exact reason. These ideas are incompatible with the mechanistic, reductionist, physicalist, and functionalist views, insofar as these are understood in mainstream scientific thought. The enactive view leads us to adopt a new vision of life, evolution, nature, and even of being and matter.

Our examination of the paradigm shift represented by the enactive approach has thus far focused on our conception of the body and of its biological nature. From a dynamical perspective, the enactive approach defines the living body as autonomous, adaptive, and existing in precarious conditions. These characteristics of the body entails that agency and development are two fundamental aspects of life and cognition. Agency is the ability of living systems to partially determine their bodily constitution and behavior, while development is the temporal unfolding of this partial self-determination as it is constrained by the specific conditions of the environment. The agential development of the body is based on norms of interaction enacted in the concrete history of living agents. The enactment of these norms is the living activity of sense-making, one that causes certain aspects of the environment to appear meaningful for agents, and that provides biases that orient their actions.

As we have seen, the enactive approach sees cognition as a form of sense-making proper to organisms with sensorimotor systems. From this standpoint, all forms of cognition are rooted in the sensorimotor interactions of a living organism and its environment. This new conception of body and cognition allows us to see cognitive agents as deeply entangled with their environments. This entanglement is primarily causal, because the constitution of the living body and its actions are determined by concrete interactions with the environment, interactions that establish processes of circular causality all along the extensive brain–body–environment system. Nonetheless, in the original project of the enactive approach, the entanglement of the body and the world goes even deeper than this. Here, the world is never disclosed to us as it is in itself, that is, as something radically separated from our understanding of it, but only by virtue of our bodily constitution and cultural history. As one of our most sophisticated cognitive and cultural activities, science is also determined by this history.

The enactive approach defines the existential condition of agents undertaking cognitive science as a fundamental circularity. The cognitive agent does not look at cognitive phenomena as if she were a neutral and disengaged observer. She looks at cognition through the lenses of her embodiment. That is, the scientist or the philosopher of cognitive science analyzes her own condition as an incarnated mind that is embedded in a sociocultural history. Nevertheless, the knowledge she acquires transforms her embodiment, and thereby changes the conceptual, material, and technological resources she works with. Consequently, doing cognitive science redefines the phenomena the agent observes, and thus allows her and other agents to create new interpretations of themselves, ultimately making new sense of their being-in-the-world.

The recognition of a fundamental circularity should help us see enactive cognition as a pluralist field, so that any intention of a unification should be treated with skepticism (Varela, 1984). For this reason, although I argued in favor of the vision of mind and life in the enactive approach over other approaches in cognitive science, I do not pretend to dismiss the importance of other forms of cognitive science, especially other enactivisms and ecological approaches. We cannot revert to a monistic and reductionist stance and commit the same mistakes as modern Western philosophy's totalitarian epistemologies. We need an enactive cognitive science that is engaged in a productive transformation of knowledge, or as Di Paolo (2020) puts it, we need productive contradictions.

For this reason, my claims and criticisms of other forms of world-involving cognition are intended to open discussions for refining and improving concepts and tools. It is not a call for the enthronement of one research program over the rest. For instance, in my account of radical enactivism as a form of weak enactivism, I pointed to the way this view still resists abandoning prejudices that obscure our understanding of life as a phenomenon involving agency from the very beginning. However, the strong arguments of radical enactivism against representationalism are highly valuable not only because they help us reject neurocentric stances, but also because its definition of information as contentless invite us to clarify and refine the concept of sense-making (see, e.g., Thompson, 2018).

Constructing new bridges and adjusting our concepts so that they attune to the diverse models and theories of world-involving cognition, and to our observations of cognitive phenomena, is also possible. For example, Rolla and Huffermann (2021) make an exciting link between the visions of radical enactivism and the enactive approach, one that sees *know-how* as the fundamental aspect of the concepts of sense-making and radical enactivism's descriptions of information. Similarly, the work of Heras-Escribano (2019) makes the parallels between Gibson's thought and Dewey's naturalism more explicit, thereby getting closer the philosophical backgrounds of ecological psychology to those of the enactive approach. I hope my insistence on recognizing phenomenology, not as a transcendental but as an existential philosophy, also help us overcome the prejudices of some cognitive scientists when it comes to this philosophical movement, and allows building stronger links between different forms of world-involving cognition.

As a philosopher influenced by phenomenology, I believe we need the phenomenon itself to guide our scientific explanations, instead of adapting conditions of observation and experimentation to suit our theoretical prejudices. Nonetheless, we must be aware of the value of each type of approximation to cognitive phenomena. They are valuable tools that can assist us to create practical solutions.

Every scientific and philosophical approach opens a new way to look at things. None of them are necessarily wrong nor false. When a neuroscientist observes the correlation of brain activity to a cognitive function, she is not wrong about disclosing the causal basis of cognition. The manipulation of phenomena in controlled environments does not imply that scientists are misleading us about the existence of a specific phenomenon. Some explanations work and are faithful to their adequate contexts. Like any other sociocultural activity, science is a situated practice; it involves relevant aspects for certain people in a determinate sociocultural context, and with specific aims. I think that the goal is to avoid any authoritarian view that aims to instantiate universal laws that work everywhere and at every time.

Similarly, we should not pretend to build a vision of the word as it is in itself. There is no world in itself apart from our situatedness (Merleau-Ponty, 1964). Instead, our situation is the world itself. Our cognitive and sensory apparatus is the result of history; it is a cultural and natural history that roots our bodies not only in our proximate niches, but also in our planet and the whole universe. Our aspirations to study cognition scientifically should not seek for universal claims. Instead, they need to recognize that we are contingent beings. Contingency is part of our world, and there is no real gap between our bodies and our world. This gap emerges only when we erroneously think there is a world in itself that is lived by nobody. A world that is studied, observed, and explained by *no body*.

Going back to the things themselves, as Husserl (1900) urged us to do, should involve going to the concrete situations where events happen, to those places where we study and explain phenomena. This is a challenge, because science and philosophy do not set out to understand each individual thing in its specific context. Instead, we must find invariants, regularities, patterns, and we must make predictions and models of the phenomena we observe. However, every time we approach a new situation, there will be new variables and things to discover. Not only does the landscape we study change, but we also change because we are embodied beings; our tools, symbolic resources, and social norms are constantly evolving. It is for this reason that the enactive approach tells us that as scientists, we live in a fundamental circularity, one where we study our own existential condition. However, this circularity is infinite because every discovery changes us, and thereby changes the world, changing us once again.

From this standpoint, I have worked to open new discussions within the field of enactive cognition. In the next section, I summarize the three main ideas that aim to transform the enactive approach so that it is even more revolutionary, particularly from an ecological perspective. Such a radical vision is required to overcome prejudices hiding our embodied nature and deep attachment to the world around us.

7.2 Emplacing Bodies

Although the entanglement of the body and the world lies at the base of our own constitution and actions, the explicit disclosure of this entanglement challenges our philosophical and scientific inquiries. The enactive approach represents one of the most significant scientific efforts to disclose this entanglement, but it still falls short of recognizing its full depth and thickness. This book has aimed to look further into the body–world entanglement and thus strengthen the claims of the enactive approach. To carry out this work, I have proposed three central ideas.

The first is the concept of *norm development*, which I used to redefine the concept of sense-making. According to its classical description, sense-making consists in the emergence of significance in a value-neutral physical world, as a result of the autonomous activity of living organisms. This description, I argued, is based on highly abstract scenarios that successfully illustrate the normative-oriented behavior of organisms, but that fall short of illustrating the full extent of sense-making. To address this, I proposed an alternative definition of sense-making as the enactment of norms of interaction within the already established set of norms that have previously entangled the body, the world, the agent, and the environment. In this sense, I defined *norm development* as the *enactment of new norms from already given norms*. By this account, sense-making is a process of norm development. Admittedly, this phenomenon is partially illustrated by the sensorimotor account of the enactive approach (sensorimotor habits). However, I have argued that the enactive approach still needs to recognize more explicitly a related idea, namely that the local environment of cognitive agents is not a mere set of physical constraints but is also and more fundamentally a *normative field* that is similar to the realm of ecological information and affordances described by ecological approaches. This leads me to a second idea: *the environment as an active ecological field.*

In defending this ecological field, I argued against typical explanations of the enactive approach that conceive the environment as nothing more than a set of physical constraints, and as a purely material environment. By contrast, I argued that the environment plays an active role in the interactional system of life and cognition, because it is a field of forces that constrains and demands actions from agents in a way that a purely material realm cannot. Significantly, these forces are constituted by norms enacted in the past of the agent–environment system and are sedimented in habits of the body and in layers of affordances located in the environment.

Understood as a normative field, the environment is fundamentally relational, as ecological approaches rightly point out. But contrary to the claims of these approaches, the norms of interaction between agents and the environment are not fully predefined by causal processes independent of individuals' autonomy and development.

Norms of sense-making are norms enacted by agents in light of the specific contingencies of the world they find. The normative domain of interactions within the agent–environment system always remains open-ended to small or almost insignificant reconfigurations. These almost insignificant reconfigurations allow agents to readapt their behavior to the world's ever-changing circumstances. The normative field itself may also be transformed into a whole new interactional domain that enables agents to increase their repertoire of bodily skills. From this perspective, the body–world relation is a dialectical tension between the inherited past and the contingent present—one that looks normatively forward to a possible future. The tension inside the system fuels its change and transformation and motivates the enactment of new norms of sense-making.

One way of describing the transformative process of the body–world entanglement proceeds from Merleau-Ponty's phenomenological description of levels. In contemporary terms, levels are ecological norms enacted in the history of interactions of an agent–environment system but are still open to new transformational processes. These levels are scaffoldings supporting the enactment of new norms. Levels are not like affordances or things, because they are not specific features of the environment that afford bodily action. Instead, levels are structural aspects of the environment that characterize the support and background in which our perception of affordances and things occur. Phenomenologically speaking, levels are horizonal aspects constituting the normative field experienced by agents.

The third important idea I have proposed suggests that sense-making, understood as the development of norms that include the environment as a normative field, consists of enacting place norms. From a phenomenological perspective, place is a multi-horizonal structure that stabilizes and makes possible the emergence of specific contents of consciousness such as things and motor significations (i.e., affordances). From an enactive standpoint, place consists of the metastability of ecological niches that embody the web of normativities relevant to agents' interactions with things and the local environment.

The enactivist notions of microworlds and participation genres are currently the closest descriptions of an enactive place we have. However, these notions are more focused on the dynamical constitution of the metastable network of interrelated processes of place than they are on the sedimented structures of this network in the material localities of agents. I have argued that the enactive and ecological descriptions of place point to the sedimented structure of material localities that disclose the normative field in the agent–environment interactions. This structure is constituted by normativities enacted at some point in the past of an agent–environment system. Although all these norms are enacted in the temporal development of agents, the normative past transcends the individual history of agents. From an external point of view, this past is found in the phylogenesis of organisms and, in the case of humans, in the cultural traditions of societies.

An enactive account of place intends to go farther than the notions of microworlds and participation genres given in the enactive perspective. As a sort of metastable structure, place has a degree of resilience that sustains agents' individual and collective habits against the tendencies of their autonomy. Places, for instance, sediment the emergent autonomy of participatory sense-making, and support a dynamical organization that can result in a heteronomous determination for agents. Places can even function as pivots or hinges for transitioning from one microworld to another. That is, places are localities for more than one microworld. At some level, they can have a unity and identity that surpasses the concreteness of one particular set of actions.

Sense-making consists of enacting place norms, because norms of interaction in the agent–environment system are constrained and motivated by the structure of place. The transformation of these norms also entails a certain degree of modification of place. The resilience of place sustains an identity and a continuity in the normative field, so much so that it is only when the normative transformation process is dramatically changed by an event that disrupts the continuity of the agent–environment system that a place transforms or even breaks its unity and identity.

Theoretically, an enactive theory of place could be a rich field of research for enactive cognition. The sketch I have offered here scarcely reveals the multiple dimensions that must be more carefully drawn from both dynamical and phenomenological perspectives. Here, it is worth mentioning that differences between sociocultural and natural places still need to be addressed. Sociocultural places have become an object of study by scholars of ecological approaches (e.g., Barker, 1968; Heft, 2018; Rietveld & Brouwers, 2017), but these ecological accounts still need to be assimilated and adapted to an enactivist framework to produce a proper study of enactive place. The question of natural places is also a subject advanced by ecological theories of development and evolution (cf. Griffiths & Stotz, 2018; Laland et al., 2000), but these theories tend to put aside any claim about the biological autonomy of agents. For all these reasons, I think that Thompson's (2007) proposal of a research project for enactive evolution should involve finding an adequate theory of natural enactive places.

In the final section of this chapter, I address some potential research projects for the concept of enactive place, and I will consider the challenge of studying enactive cognition in place. In what follows, I intend to clarify certain aspects of the three ideas central to this book.

7.3 Situating and Enlightening

The three main ideas I have proposed to help understand the emplacement of bodies from the standpoint of the enactive approach need to be clarified to avoid misunderstandings of my original aim.

First, my criticism of the theory of significance as a surplus does not pretend to minimize the value of the conceptual framework of the enactive approach, one that led the founders of this approach to craft the very notion of sense-making. The enactive approach inherits many ideas from second-order cybernetics that ground the systemic view of the enactive approach. Within this paradigm, it seems reasonable to conclude that cognition and life are based on the co-emergence of a self and of an Umwelt. The founders of the enactive approach look for the systemic requirements of the constitution of autonomous systems and the sort of interactions they maintain with their surroundings. Be this as it may, cybernetics is a highly abstract science that can miss what phenomenology reveals about cognitive phenomena.

Phenomenology is a science of experience and seeks to disclose invariants of how we relate to the things around us, and to ourselves. Although phenomenological analysis is not infallible, and many abstractions occur in its unveiling of the basal levels of our embodiment, this philosophical movement tries to avoid the theoretical prejudices that usually obscure our scientific observations. Following the principles of mutual enlightenment (Sect. 4.1.3) and of fundamental circularity (Sect. 3.1.2), we should not aspire to any abstract enclosure of what phenomenology reveals about the richness of cognitive phenomena. Neither should we aspire to scientifically model all the complexity phenomenology discloses when it comes to cognitive phenomena, either. Instead, we need to find paths that balance our theoretical constructions with our practical concerns.

Consequently, my call for a shift in the definition of sense-making does not imply a rejection and replacement of its original meaning. As it is, this definition helps us understand some aspects of enactive cognition and sense-making. It also serves to create models of cognition that visualize the synchronic constraints at play in the relation of an agent and the environment. However, as I have shown, there are cases in which this view has limits, and we cannot pretend that understanding sense-making as simply the emergence of norms from raw materiality is the most reliable description of sense-making. Norm development is not only a more accurate description of sense-making from a theoretical and phenomenologically based point of view. Norm development unveils the ecological and historical dimensions of sense-making, and it is sometimes central to look at these dimensions for practical purposes.

Think, for example, about psychiatry and psychopathology. In these fields, we want to help people to overcome pathologies that cause significant alterations to their emotional and cognitive lives. To do this, we can produce explanations and models at different levels of intervention. Fuchs (2018) proposes we look at two different levels. First, at the level of brain processes, we can analyze and intervene in the physicochemical interactions of the brain. Interventions at this level can offer partial solutions that enable other treatments like psychotherapy. At the level of psychotherapy, analyses and interventions concern the constitution of the whole person and her multiple dimensions of embodiment. Altering interpersonal relationships and emotional responses can help heal people in grief.

Nonetheless, this non-reductionist vision of psychiatry might still be insufficient in some specific cases, particularly when we need to understand the origin of a mental pathology in the community where a person lives (Ferrari et al., 2013). Depression, for instance, has become one of the most critical mental health problems in contemporary Western civilization. There could be many factors at play for this increase in pathological cases, and as a result, it is worth looking at the sociocultural realities of our societies for conditions that foster the emergence of depression. To what extent do our interactions with others—mediated as they are by the capitalist system of production—cause the loneliness, anxiety, and desperation we see in depressed people? How much does the consumer society encourage people's permanent dissatisfaction with their lives? How do our modern cities result in a distancing between people that does not exist in smaller, and more traditional communities?

Many of the responses to these questions involve more exhaustive analysis than those proposed by Fuchs. Once again, the point is not to be unfair with the value of the two approaches that Fuchs suggests. The fact is that we sometimes need a wider scope to study phenomena like psychopathologies. The ecological and historical dimensions of place might be helpful for such a study, since these dimensions involve the interrelation of one subject to others, in specific locations and within the historical constitution of a network of matter and meaning (Ratcliffe, 2014).

In sum, the view of enactive place and norm development does not aim to replace the original framework of the enactive approach but seeks instead to open an alternative and coexistent view of sense-making, one that is useful for theoretically and practically enriching the labor of enactive cognitive scientists.

When we turn to the claim that the environment should be defined as an active field of ecological forces, it is important to note that such an idea does not deny that we can sometimes model the environment as a mere set of physical constraints. Di Paolo et al. (2014) achieved this when they created models of sensorimotor contingencies based on Piaget's theory of equilibration. The strength of the ecological view is that it can help us to understand why some developmental paths are more suitable than others. Such an understanding is not possible by a simple appeal to a model based on an agent's autonomy, and to the set of material constraints in the environment.

Material constraints are not enough to limit an agent's capacity to enact some specific sense-making norms instead of others. Materiality offers infinite possibilities to our bodies, which are organically, dynamically, and interactively highly flexible. Despite this flexibility and openness, we tend to preserve patterns of behavior and development. Normative constraints that are inherited and transmitted by other agents may do a better job of explaining the regularity of such patterns, and Barker has already showed this in his ecological approach to sociocultural places (Sect. 6.2).

In theoretical biology, Raimondi's description of an epigenic field for ontogenesis significantly coincides with my view of the environment as an active ecological field. In ontogenesis, there are "internal" structures of the organism (e.g., the genome) that predisposes the expression of a determinate phenotype (i.e., the agent's body). There are also relevant environmental structures that causally contribute to the constitution of the phenotype (the ontogenic niche). The epigenic field comprises the relational space between internal and external structures that serve to constrain an organism's "developmental pathways" (see also Rolla & Figueiredo, 2021). The field must not be understood as the bare material landscape (the ontogenic niche) that causally interacts with the body's materiality (the phenotype). Instead, the field is the domain where the process of the agent–environment system development constrains itself, determining the conditions of its continuity, and thereby establishing its limits and potential for change. The existence of such a field, therefore, limits the possible number of interactions between the agent and the environment, and can explain why some development pathways are taken instead of others.

We can infer the same idea from the work of de Carvalho and Rolla (2020), who define ecological information as environmental structures determined by processes of reduction of uncertainty. These authors think that cognition consists of processes of learning bodily skills or of acquiring different forms of know-how (see also Rolla & Huffermann, 2021). Like radical enactivism, de Carvalho and Rolla define information as covariation between worldly events that reduce uncertainty. The reduction of uncertainty limits the number of environmental aspects that are relevant to the agent's action. Since learning new skills determines the number of possible actions, when a new environmental circumstance appears, the agent will develop new skills, but is limited to the number of possible actions she has previously learned. Hence, further information is determined by the previously acquired information. Consequently, ecological information becomes a relational field that limits the possibilities of an agent's development and the ecological field's unfolding.

Further work is needed to connect all these ideas suggesting the existence of a relational field that predisposes the development of the agent and its environmental counterpart. However, the need for an ecological description of the environment like the one I have made seems to be a growing necessity for the enactive approach.

Considering the notion of place, it should not be confused with other similar accounts of the ecological realm based on refinements of the modern synthesis of evolutionary biology. What has been referred to as *the extended evolutionary synthesis* recognizes ecological processes as fundamental aspects of natural and cultural evolution (Laland et al., 2015). The ideas of niche construction theory are central to understanding the causal intertwining in processes of evolution and have already inspired many works in the field of world-involving cognition (e.g., Bruineberg et al., 2018; Clark, 2008; Heras-Escribano & de Jesus, 2018; Hutto & Myin, 2017; Sterelny, 2010). However, the account of place in the enactive and phenomenological perspectives I have proposed goes far beyond ideas available in evolutionary biology.

The niche construction theory describes a mutual co-determination of organisms and environment that can be put in terms of dispositionalism (Heras-Escribano, 2019), or as a statistical covariation of the internal states of organisms and the features of the environment (Bruineberg et al., 2018). I am pointing here to something more than what is contained in these descriptions. I am arguing that the entanglement of the body and the world generates a normative structure in the environment that is not only material or statistical, but also meaningful. As an enactive view, I endorse the claim that meaning depends on the constitution of an autonomous organization. If we look only at this organization and its structures, I do not think we will find the origin of the norms agents enact while interacting with their environments. The ecological intertwining of agents with multiple forces at play in their niches also determines the development and evolution of agents' bodies and behavior.

I find similarities with my description of place in Werner's (2020) idea of a *cognitive niche*. He describes the process of enacting a world—as described by Varela and colleagues in *The Embodied Mind*—as a process of niche construction. Yet his account does not refer to the material niche construction many other researchers of world-involving cognition point to. Instead, Werner describes the process of niche construction as the active transformation of the environment relevant to an agent, and in light of the autonomy of the agent.

The increasing number of proposals seeking explanations for the attachment of the living body to the environment from the perspective of the enactive approach gives us some idea as to the necessity of opening another axis of enactive cognition to the study of life and mind in place. Such a study will be challenging nonetheless, and we may need to work on ontological, epistemological, and methodical questions to produce practical and operational modeling of enactive places. This being the case, the last question I will address in this book is the study of bodies in place, because here, the way is open and full of promise, despite significant challenges.

7.4 Studying Bodies in Place

The concept of place is vast and rich, and it can be difficult to choose a case that exemplifies all the dynamics at issue when we use the term from an enactive-ecological perspective within cognitive science. Nevertheless, I think the example of Mexican chinampas is paradigmatic, because it shows clearly both the ecological and dynamical relations of a place. The human agents responsible for the construction of chinampas are aiming to sustain the ecological balance of place, and this purpose significantly guides their way of life. Nonetheless, it is also important to note that places are not only places of dwelling in the literal sense. The example of *Chinampas* is a concrete expression of an extended notion of place, and this by virtue of their unique materiality.

There are also more virtual enactive places. Merleau-Ponty (2012), for instance, described the emergence of a musical dimension of space when skilled musicians play. This musical space can fulfill the role of a more virtual structure that helps agents to share a communal ground for habitual patterns of interaction (e.g., by dancing). Only those agents familiarized with this musical place can inhabit the virtual space and smoothly interact with others *in* it. The case of social media is another example of this more virtual kind of place.

There is also a significant affective dimension of place. Some diagnoses of depression link the pathology to a feeling of *having* no place. This should not be understood metaphorically. Place is not in itself; it is the intertwining of every embodied being to its surroundings, and this intertwining is significantly affective. When the study of the psychopathology of depression adopts a phenomenological stance, it reveals that affectivity, emotion, and empathy are constitutive aspects of our experience of the world as a place of dwelling (Ratcliffe, 2014). The negative alteration of our world experience, understood as a place we share with others, diminishes our abilities to perceive natural and social affordances. This alteration should not be seen as a merely subjective or even as a bodily disability, but an abnormal reconfiguration of the agent–environment system where the process of norm development seems to be wasted (Toro et al., 2020). That would prove that the constitution of the enactive place needs the capacity of the agent–environment system to actualize and adapt the normative reconfiguration of the system to the new tensions that emerge from the system's activity. Otherwise, places, including the world as a place of dwelling, lose their role of anchoring the living body in a domain shared with other agents (Jacobson, 2009). Such a study also reveals the affective dimension of place and the ecological character of emotions.

Any study of the richness and complexity of place from an enactivist perspective faces the challenge of meeting the standards of cognitive science. Admittedly, I constructed my approach to place from a holistic standpoint that might not fit easily within the current practices of this branch of science. We must address this challenge if we wish to produce operative and predictive models of cognition.

While operative frameworks may not directly model the holistic description of place that I defend, I think they can at least establish a theoretical framework against which more simple models can be created. For instance, Di Paolo et al. (2018) propose that concepts like sense-making, autonomy, and agency should work as heuristic resources. Therefore, a general and abstract definition of these concepts can guide research in concrete fields of science (e.g., psychiatry, developmental phycology, musicology, etc.) and adapt to those fields' more concrete aspects. More specific applications of the general concepts can, in return, give us feedback from concrete case studies, and force us to reaccommodate and refine our general concepts. I believe the same strategy can work for an enactive account of place. A broader definition of place can work as a guide for our research in multiple fields without requiring us to model all the dimensions of place at once. In exchange, concrete results in a given field can enrich and expand a general theory of enactive place.

The enactive theory of place can also play a significant part in helping us reconsider the type of scientific endeavor pursued by enactive cognition. Cognitive science has traditionally sought to produce highly formal, measurable, and predictive models of cognition. However, this is not the only way to carry out science. Hendriks-Jansen (1996), for instance, recovering Nagel's (1961) categorization of the four types of scientific explanation, writes that cognitive science has indeed attained formal-universal, probabilistic, and teleological-functional explanations of cognition, but has failed to attain more genetic explanations.

The value of genetic explanations is that they do not aim to state universal laws that can explain phenomena once and for all. Nor are such explanations intended to produce inferences based on probabilistic models or explain the behavior of a system based on its mechanical functionality. A genetic explanation looks for the historical origins of a phenomenon, thereby unveiling its leading causes. One of the advantages of this kind of explanation is that it tends to focus less on the abstract statements and general theories that science produces, and more on the detailed and careful observations of a phenomenon as it occurs in the field.

Situating our study of enactive cognition in place from such a perspective would more closely resemble the work of anthropologists, ethnologists, and archaeologists. These fields understand human agency in the wild, as it were, using mainly abductive types of reasoning. However, it is essential to note that qualitative analyses and genetic explanations are not exclusive to the humanities. Hendrick Jansen, for example, cites Darwin's theory of evolution as precisely this type of genetic explanation. Nevertheless, Darwin was indeed a biological scientist who observed phenomena in the field, and his work unveiled the crucial ecological relations of the landscapes he observed, showing processes that might originate within these ecological fields.

Adopting such explanations in enactive cognition does not mean that formal and predictive models are not helpful for our scientific approaches to cognitive phenomena. The complementarity, dialogue, and mutual enlightenment of different levels and types of explanations can guide our understanding of the tremendously complex phenomenon of cognition and illuminate its underlying existential structures of the being-in-the-world.

References

Barker, R. G. (1968). *Ecological psychology: Concepts and methods for studying the environment of human behavior*. Stanford University Press.

Bruineberg, J., Rietveld, E., Parr, T., van Maanen, L., & Friston, K. J. (2018). Free-energy minimization in joint agent-environment systems: A niche construction perspective. *Journal of Theoretical Biology, 455*, 161–178.

Clark, A. (2008). *Supersizing the mind: Embodiment, action, and cognitive extension*. Oxford University Press.

de Carvalho, E. M., & Rolla, G. (2020). An enactive-ecological approach to information and uncertainty. *Frontiers in Psychology, 11*, 588.

Di Paolo, E. A. (2020). Enactive becoming. *Phenomenology and the Cognitive Sciences, 20*, 1–27.

Di Paolo, E. A., Barandiaran, X. E., Beaton, M., & Buhrmann, T. (2014). Learning to perceive in the sensorimotor approach: Piaget's theory of equilibration interpreted dynamically. *Frontiers in Human Neuroscience, 8*, 551.

Di Paolo, E. A., Cuffari, E., & De Jaegher, H. (2018). *Linguistic bodies: The continuity between life and language*. MIT Press.

Ferrari, A. J., Somerville, A. J., Baxter, A. J., Norman, R., Patten, S. B., Vos, T., & Whiteford, H. (2013). Global variation in the prevalence and incidence of major depressive disorder: A systematic review of the epidemiological literature. *Psychological Medicine, 43*(3), 471–481.

Fuchs, T. (2018). *Ecology of the brain: The phenomenology and biology of the embodied mind.* Oxford University Press.

Griffiths, P. E., & Stotz, K. (2018). Developmental Sytems theory as a process theory. In J. Dupré & D. J. Nicholson (Eds.), *Everything flows: Towards a processual philosophy of biology* (pp. 225–245). Oxford University Press.

Heft, H. (2018). Places: Widening the scope of an ecological approach to perception–action with an emphasis on child development. *Ecological Psychology, 30*(1), 99–123.

Hendriks-Jansen, H. (1996). *Catching ourselves in the act: Situated activity, interactive emergence, evolution, and human thought.* MIT Press.

Heras-Escribano, M. (2019). *The philosophy of affordances.* Springer International Publishing.

Heras-Escribano, M., & de Jesus, P. (2018). Biosemiotics, the extended synthesis, and ecological information: Making sense of the organism-environment relation at the cognitive level. *Biosemiotics, 11*(2), 245–262.

Husserl, E. (1900). *Logical investigations.* (J. N. Findlay, Trans., D. Moran, Ed.) (Vol. II). Taylor & Francis.

Hutto, D. D., & Myin, E. (2017). *Evolving enactivism: Basic minds meet content.* MIT press.

Jacobson, K. (2009). A developed nature: A phenomenological account of the experience of home. *Continental Philosophy Review, 42*(3), 355–373.

Laland, K. N., Odling-Smee, F. J., & Feldman, M. W. (2000). Niche construction, biological evolution, and cultural change. *Behavioral and Brain Sciences, 23*(1), 131–146.

Laland, K. N., Uller, T., Feldman, M. W., Sterelny, K., Müller, G. B., Moczek, A., et al. (2015). The extended evolutionary synthesis: Its structure, assumptions and predictions. *Proceedings of the Royal Society B: Biological Sciences, 282*(1813), 20151019.

Merleau-Ponty, M. (1964). *Le visible et l'invisible: suivi de notes de travail. par Maurice Merleau-Pont. Texte établi par Claude Lefort.* Gallimard.

Merleau-Ponty, M. (2012). *Phenomenology of perception.* (D. Landes, Trans.). Routledge.

Nagel, E. (1961). *The structure of science: Problems in the logic of scientific explanation.* Harcourt, Brace & World.

Ratcliffe, M. (2014). *Experiences of depression: A study in phenomenology.* OUP Oxford.

Rietveld, E., & Brouwers, A. A. (2017). Optimal grip on affordances in architectural design practices: An ethnography. *Phenomenology and the Cognitive Sciences, 16*(3), 545–564.

Rolla, G., & Figueiredo, N. (2021). Bringing forth a world, literally. *Phenomenology and the Cognitive Sciences*, 1–23.

Rolla, G., & Huffermann, J. (2021). Converging enactivisms: Radical enactivism meets linguistic bodies. *Adaptive Behavior, 30*, 345.

Sterelny, K. (2010). Minds: Extended or scaffolded? *Phenomenology and the Cognitive Sciences, 9*(4), 465–481.

Thompson, E. (2007). *Mind in life: Biology, phenomenology, and the sciences of mind*. Harvard University Press.

Thompson, E. (2018). *Review of "Evolving enactivism: Basic minds meet content"*. MIT Press.

Toro, J., Kiverstein, J., & Rietveld, E. (2020). The ecological-enactive model of disability: Why disability does not entail pathological embodiment. *Frontiers in Psychology, 11*, 1162. https://doi.org/10.3389/fpsyg.2020.01162

Varela, F. J. (1984). The creative circle: Sketches on the natural history of circularity. In P. Watzlawick (Ed.), *The invented reality: How do we know what we believe we know* (pp. 309–323). Norton.

Werner, K. (2020). Enactment and construction of the cognitive niche: Toward an ontology of the mind-world connection. *Synthese, 197*(3), 1313–1341. https://doi.org/10.1007/s11229-018-1756-1

Index[1]

[1] Note: Page numbers followed by 'n' refer to notes.

M. A. Sepúlveda-Pedro, *Enactive Cognition in Place*, New Directions in Philosophy and Cognitive Science,
https://doi.org/10.1007/978-3-031-20282-7

FSC
www.fsc.org
MIX
Papier | Fördert gute Waldnutzung
FSC® C083411

Kolibri 360 GmbH
Ferdinand-Jühlke-Straße 7
99095 Erfurt, Deutschland
produktsicherheit@kolibri360.de